EMERGING

TECHNOLOGIES IN

OIL AND GAS INDUSTRY

Matthew N. O. Sadiku, Ph.D., P.E.

Regents Professor Emeritus and IEEE Life Fellow

Prairie View A&M University

Prairie View, TX 77446

Email: sadiku@ieee.org

Web: www.matthew-sadiku.com

DEDICATION

DEDICATED TO MY WIFE JANET

FOR BEING A WONDERFUL GIFT

PREFACE

The oil and gas industry is one of the largest and most complex, diverse industries in the world, involving the exploration, extraction, refining, and distribution of hydrocarbon resources. It divides into upstream, midstream, and downstream. The industry serves as a lifeline for the global economy, powering key sectors such as transportation, electricity generation, heating, and manufacturing. In today's fast-paced digital landscape, the oil and gas industry is undergoing a profound transformation to stay competitive globally. Now, all companies in the industry use the latest technology to improve their processes, increase productivity, and protect their market. These include companies such as Shell, AML3D, ExxonMobil, Chevron, BP, Siemens Energy, Halliburton, Total, and General Electric. Companies in the oil and gas industry influence the global economy by engaging in the exploration, extraction, refining, and transportation of one of the primary fuel sources.

Technology impacts every aspect of our modern society. It is constantly changing, and oil and gas (O&G) companies must keep up to stay competitive. Natural gas and oil companies would not be successful at finding oil and gas deposits without technology. The O&G sector is undergoing rapid and significant transformation. Digital transformation of oil and gas companies is the integration of emerging technologies like artificial intelligence, cloud computing, Internet of things, big data, and data analytics across business functions. The goals of digitalization include improving efficiency, driving employee productivity, promoting sustainability, mitigating risks, and providing better customer and partner experiences.

This book explores the emerging technologies in the oil and gas sector. The book is organized into ten chapters that address these emerging technologies: artificial intelligence, big data, the Internet of things, cloud computing, blockchain, nanotechnology, 3D printing, drones, and cybersecurity. It describes each technology, its applications in oil and gas, its benefits, and its challenges. These emerging technologies can play a vital role in boosting the operational efficiency of the oil and gas industry. They demand attention within the oil and gas sector.

Chapter 1: Introduction: This chapter explores the use of emerging technologies in the oil and gas industry and serves as an introduction to the entire book. Emerging technology is a term generally used to describe new technology. This chapter briefly covers some emerging technologies such as artificial intelligence, robotic automation, cloud computing,

Internet of things, blockchain, nanotechnology, 3D printing, drones, augmented reality, virtual reality, big data, and data analytics. The new technologies in the oil and natural gas sector have enabled the explosion of production growth in the United States.

Chapter 2: Artificial Intelligence in Oil & Gas: This chapter discusses in detail and demonstrates how AI is transforming the oil and gas sector. Artificial intelligence (AI) refers to the ability of a computer system to perform human tasks (such as thinking and learning) that usually can only be accomplished using human intelligence. The impact of AI on all spheres of human life is hard to miss. Artificial intelligence, as the most important general-purpose technology of today, is rapidly transforming the oil and gas industry. Many companies are actively working with AI in the oil and gas industry, developing effective solutions. The integration of artificial intelligence into the oil and gas industry represents a pivotal shift towards addressing the complex challenges of today's energy landscape.

Chapter 3: Big Data in Oil & Gas: This chapter reviews the utilization of big data and data analytics in the oil and gas industry. Big data applies to data sets of extreme size (e.g. exabytes, zettabytes) which are beyond the capability of the commonly used software tools. There are ample opportunities for oil and gas companies to use big data to get more oil and gas out of hydrocarbon reservoirs, reduce capital and operational expenses, increase the speed and accuracy of investment decisions, and improve health and safety while mitigating environmental risks. The integration of advanced analytics and the increasing reliance on big data are driving the oil and gas sector toward a future where fully autonomous control systems for complex processing facilities become a reality.

Chapter 4: Internet of things in Oil & Gas: The objective of this chapter is to provide an informative guide on how the Internet of things (IoT) is revolutionizing oil and gas exploration. Today, the Internet has become an indispensable part of life. When it comes to the Internet, the Internet of things (IoT) has taken center stage. The Internet of things (IoT) is a concept that refers to the interconnectedness of everyday objects and devices through the Internet, allowing them to collect, exchange, and analyze data. In oil and gas exploration and production, IoT sensors are deployed on drilling equipment and pipelines to monitor parameters such as pressure, temperature, and flow rates. Today, the industrial Internet of things (IIoT) is the future of the oil and gas industry.

engineering, and the use of materials at the nanoscale. It has stood strong in the oil and gas industry, with many applications that have gone from laboratory and simulation studies to successful trial applications in the field. The oil and gas field is expected to see a successful expansion in the application of nanotechnology in different areas. It will continue to influence the future of the oil and gas sector by embracing ongoing research, development, and responsible application.

Chapter 8: 3D Printing in Oil & Gas: This chapter aims to highlight significant opportunities and challenges relating to the adoption of 3D printing in the oil and gas industry. Additive manufacturing (or 3D printing) is a valuable tool that enables the faster and more efficient manufacturing of spare parts, equipment, and components while reducing downtime and maintenance costs. To reduce wasteful consumption and accidents due to leaking, oil & gas companies are turning to 3D printing to create geometrically complex, cost-effective parts. Major companies across the oil and gas industry have adopted industrial-scale additive manufacturing capabilities for production.

Chapter 9: Drones in Oil & Gas: This chapter examines the applications of drones in the oil and gas industry. The FAA defines drones, also known as unmanned aerial vehicles

(UAVs), as any aircraft system without a flight crew onboard. The application of drone technology has transformed many industries, including the oil and gas industry. Drones in oil and gas have helped reduce inspection time, cut costs, decrease downtime, and identify problems early on. The adoption of drone technology will become increasingly common throughout the oil and gas industry as companies discover the advantages of using drones. When it comes to oil and gas industry, drones are a super significant investment.

Chapter 10: Cybersecurity in Oil & Gas: The chapter focuses on systematically exploring cybersecurity and safety challenges of the O&G sector. Cybersecurity refers to a set of technologies and practices designed to protect networks and information from damage or unauthorized access. The oil and gas industry is a prime target for cyber-attacks due to the high value of the data and systems they control. Cyber-attacks on the oil and gas industry are growing because the sector is becoming increasingly dependent on technology and automation. Oil and gas companies that maintain cybersecurity as a central tenet of their digital strategy stand to gain the most. Emphasizing cybersecurity today means a more secure and resilient oil and gas infrastructure tomorrow.

This is a comprehensive book on emerging technologies in oil and gas industry. It provides an overview of each emerging technology, its application, benefits, and challenges in oil and gas. It does that in a way that beginners can easily understand them. It is a must-read book for those interested in the new technologies in the oil and gas industry, which plays a crucial role in a functional, modern society.

I am grateful for the support of Dr. Annamalia Annamalai, the department head of the Department of Electrical and Computer Engineering, and Dr. Pamela Obiomon, the dean of the College of Engineering at Prairie View A&M University, Prairie View, Texas. Special thanks are due to my wife Dr. Janet Sadiku for helping in various ways. This book is dedicated to her.

- M. N. O. Sadiku

ACKNOWLEDGMENT

I am grateful for the support of Dr. Annamalia Annamalai, the department head of the Department of Electrical and Computer Engineering, and Dr. Pamela Obiomon, the dean of the College of Engineering at Prairie View A&M University, Prairie View, Texas. Special thanks are due to my wife Dr. Janet Sadiku for helping in various ways. To her, this book is dedicated.

- M. N. O. Sadiku

ABOUT THE AUTHOR

Matthew N. O. Sadiku received his B. Sc. degree in 1978 from Ahmadu Bello University, Zaria, Nigeria, and his M.Sc. and Ph.D. degrees from Tennessee Technological University, Cookeville, TN in 1982 and 1984 respectively. From 1984 to 1988, he was an assistant professor at Florida Atlantic University, Boca Raton, FL, where he did graduate work in computer science. In total, he received seven college degrees. From 1988 to 2000, he was at Temple University, Philadelphia, PA, where he became a full professor. From 2000 to 2002, he was with Lucent/Avaya, Holmdel, NJ as a system engineer and with Boeing Satellite Systems, Los Angeles, CA as a senior scientist. He is presently a Regents professor emeritus of electrical and computer engineering at Prairie View A&M University, Prairie View, TX.

He is the author of over 1,290 professional papers and over 140 books including "Elements of Electromagnetics" (Oxford University Press, 7th ed., 2018), "Fundamentals of Electric Circuits" (McGraw-Hill, 7th ed., 2020, with C. Alexander), "Computational Electromagnetics with MATLAB" (CRC Press, 4th ed., 2019), "Principles of Modern Communication Systems" (Cambridge University Press, 2017, with S. O. Agbo), and "Emerging Internet-based

Technologies" (CRC Press, 2019). In addition to the engineering books, he has written Christian books including "Secrets of Successful Marriages," "How to Discover God's Will for Your Life," and commentaries on all the books of the New Testament Bible. Some of his books have been translated into French, Korean, Chinese (and Chinese Long Form in Taiwan), Italian, Portuguese, Spanish, German, Dutch, Polish, and Russian.

He was the recipient of the 2000 McGraw-Hill/Jacob Millman Award for outstanding contributions in the field of electrical engineering. He was also the recipient of Regents Professor award for 2012-2013 by the Texas A&M University System. He is a registered professional engineer and a life fellow of the Institute of Electrical and Electronics Engineers (IEEE) "for contributions to computational electromagnetics and engineering education." He was the IEEE Region 2 Student Activities Committee Chairman. He was an associate editor for IEEE Transactions on Education. He is also a member of Association for Computing Machinery (ACM). His current research interests are in the areas of computational electromagnetic, computer science/networks, engineering education, and marriage counseling. His works can be found in his autobiography, "My Life and Work" (Author's Tranquility Press, 2024) or on his website: **www.matthew-sadiku.com.** He currently resides with his wife Janet in

Westlake Florida. He can be reached via email at **sadiku@ieee.org**

CONTENTS

CHAPTER 1

INTRODUCTION

"Always expect the unexpected. The oil and gas industry is terrible at predicting anything. Always have a back-up plan."

– David Dixon

1.1 INTRODUCTION

For over a century, oil and gas has played a vital role in the economic transformation of the world. The oil and natural gas industry provides the world with reliable and affordable energy. The industry is the largest and most diverse one, with operations spanning a wide range of conditions in several countries. It is illustrated in Figure 1.1 [1].

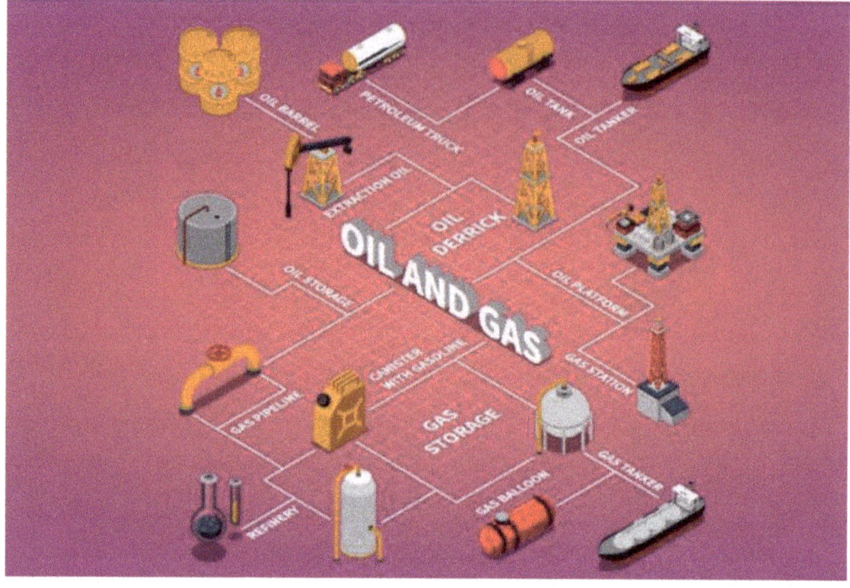

Figure 1.1 Representation of the oil and gas [1].

It includes large networks of equipment sourcing, trading, and transport. It uses cloud computing, data analytics, artificial intelligence, and machine learning to optimize production. Oil and natural gas technologies have dramatically altered the energy landscape over the past decade, particularly in North America. The industry is known for its complex infrastructure, as typically shown in Figure 1.2 [2].

Figure 1.2 The O&G industry is known for its complex infrastructure [2].

Technology is constantly changing, and oil and gas (O&G) companies must keep up to stay competitive. The oil and gas industry has always been at the forefront of technological advancements, constantly pushing the boundaries of innovation to enhance efficiency, safety, and sustainability. The O&G sector is undergoing rapid and significant transformation. Digital transformation of oil and gas companies is the integration of emerging technologies like cloud services, automation, IoT, big data, and data analytics across business functions.

In the fast-paced world of oil and natural gas industry, staying ahead means embracing the new. To boost efficiency, cut costs, and lessen environmental harm, the oil and gas industry is adopting new technologies. This is becoming more achievable with the rise of emerging technologies like artificial intelligence, machine learning, robotic automaton, big data analytics, the Internet of things, and drones. Emerging technologies are finding their way into the oil, gas, and petrochemical industry where they provide a wide range of benefits. The integration of these technologies into the oil and gas sector is a strategic move towards a more sustainable, safer, and cost-effective future [3].

This chapter explores the use of emerging technologies in the oil and gas industry. It serves as an introduction to the entire book. It begins with describing what emerging technologies are all about. It briefly presents some emerging technologies in the oil and gas industry. It explains the oil and gas industry. It highlights the benefits and challenges of emerging technologies in the oil and gas industry. It concludes with comments.

1.2 WHAT ARE EMERGING TECHNOLOGIES?

Technology may be regarded as a collection of systems designed to perform some function. It can help alleviate some of the challenges facing business today. Emerging technology is a term generally used to describe new technology. The term often refers to technologies currently developing or expected to be available within the next five to ten years. Any imminent, but not fully realized, technological innovations will have some impact on the status quo.

Emerging technologies are shaping our societies. They continue to affect the way we live, work, and interact with one another. Emerging technology (ET) lacks a consensus on what classifies them as "emergent." It is a relative term because one may see a technology as emerging and others may not see it the same way. It is a term that is often used to describe a new technology. A technology is still emerging if it is not yet a "must-have" [4]. An emerging technology is one that holds the promise of creating a new economic engine and is trans-industrial. ET is used in different areas such as media, healthcare, business, science, education, or defense.

The characteristics of emerging technologies include the following [5]:

- *Novelty:* Emerging technologies are typically new or novel, meaning they have yet to be widely adopted or used. They often represent a significant departure from existing technologies or processes.

- *Potential for Disruption:* Emerging technologies have the potential to disrupt existing markets, industries, or ways of doing things. They may also displace existing businesses or industries.

- *Uncertainty:* Because emerging technologies are still in the early stages of development, there is often a high uncertainty surrounding their future potential and impact. It can be challenging to predict how they will evolve.

- *Rapid Change:* Emerging technologies often evolve rapidly, with new developments ants and innovations emerging frequently. It can make keeping up with the latest trends and advancements challenging.

- *Interdisciplinary*: Emerging technologies often involve multiple disciplines or fields of study, such as computer science, engineering, and biology. They may require collaboration across different fields and industries to develop their potential fully.

Emerging technologies are worth investigating. They are responsible for developing new products or devices. The military often looks to emerging technologies for new services or tools that will help them create a competitive business advantage.

1.3 OIL AND GAS INDUSTRY

The oil and gas industry is complex and diverse. The industry deals with lots of data coming from manufacturing processes. It divides into upstream, midstream, and downstream, as shown in Figure 1.3 [6].

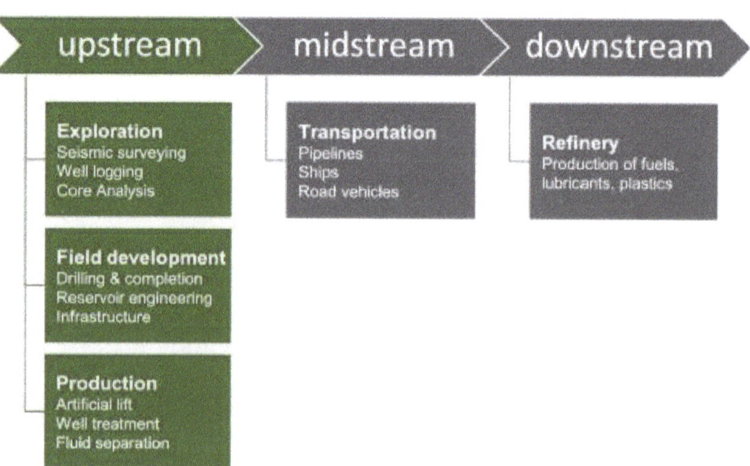

Figure 1.3 The oil and gas industry divides into upstream, midstream, and downstream [6].

The upstream summarizes the subsurface (mining) part of the industry, including exploration, field development, and production of crude oil/gas. The upstream segment is the most capital-intensive part of oil and gas and the segment of enormous uncertainties to tackle. It covers crude oil and natural gas production. It includes searching for potential underground or underwater crude oil and natural gas fields, drilling exploratory wells, and subsequently drilling and operating the wells used to lift the crude oil or raw natural gas to the surface. Companies from the sector deal with enormous uncertainties handled manually and relied on expert knowledge, not the actual data.

Midstream stands for storage and transportation of oil and gas. In midstream business, AI can support proper planning and execution, optimal route selection, etc. In contrast, it helps refiners plan optimal blending, forecasting the demand, estimating prices, and improvising customer relationships in the downstream business.

Downstream is for refinery and distribution, i.e., production of fuels, lubricants, plastics, and other products. By embracing AI, downstream firms achieve refinery and distribution cost reduction and reach regulatory compliance faster in several ways. Beyond crude oil and natural gas, downstream oil and gas companies benefit from AI-driven predictions of consumer demand for various products. The application of AI in downstream oil and gas companies involves real-time monitoring systems that optimize refineries. Those systems monitor operations and collect data during distillation, catalytic cracking, and hydrogenation. They also screen data from energy meters and equipment sensors. Downstream operations can be optimized to minimize costs and maximize spreads. AI predictive maintenance reduces AI downstream processes and optimizes maintenance planning.

1.4 EMERGING TECHNOLOGIES IN OIL & GAS

As the energy demand continues to grow, the oil and gas companies have embraced various technological advancements to extract, refine, and distribute oil and natural gas resources more effectively. These technological advancements have revolutionized the O&G industry. Emerging technologies used in the oil and gas industry include the following [7-9]:

- *Artificial Intelligence* (AI): This is changing the game across various industry facets. AI provides powerful benefits across the value chain in energy production. Oil and gas companies use the technology to assess the value of reservoirs and customize drilling plans according to an area's geology. AI is also used during location searches for oil drilling to analyze seismic data and provide risk insights. By applying AI, firms can now evaluate oil reservoirs with higher precision, monitor machinery conditions in real-time, and even devise drilling plans tailored to specific geological conditions. This leads to enhanced operational efficiency, reduced chances of human error, and significant cost reductions. The oil and gas industry is also embracing the power of artificial intelligence and data science to tackle intricate challenges across all stages of their operations. Artificial Intelligence's primary goal in the oil and gas upstream, midstream, and downstream operations is to reduce risks, improve efficiency, and extract more value from resources by using advanced technology.

- *Machine Learning*: The oil and gas sector is undergoing a rapid transformation with the introduction of machine learning, an emerging field that has the potential to transform exploration and production methodologies. This tool gives oil and gas companies powerful insights from the data generated during oil and gas operations. They can use the data to improve operational efficiency and decision-making. Machine learning models look at sensors to help you find opportunities for improvement. Machine learning algorithms excel at analyzing production data and identifying patterns that, when used, optimize performance, resulting in increased productivity and reduced downtime.

- *Robotic Automation*: Engineers have found ways to automate tasks and overcome sub-surface challenges using the latest technology. Emerging technologies in robotics and automation have the potential to improve operations in the oil and gas industry by reducing costs and increasing safety, efficiency, and speed of the processes. One of the highly adopted automation technologies is robotic process automation (RPA). The technology automates repetitive and rule-based tasks such as document processing, data entry, and reporting. When an oil and gas well is being closed, RPA significantly reduces closing time and minimizes human errors. Robotic systems can perform tasks such as pipeline inspection, maintenance, and repairs, minimizing the need for human intervention in potentially dangerous situations. Most of the robots designed for the oil, gas, and petrochemical industry can withstand extreme weather, temperatures and other harsh conditions usually present in the oil and gas industry.

• *Big Data and Analytics*: This is another tool in the tech arsenal that is making waves. It is all about using the vast amounts of data generated daily to make smarter decisions. The oil and gas industries generate large volumes of data every day. Using analytics, data scientists can find trends and patterns in the data. The oil and gas industry has been collecting data for decades, so it is no surprise that companies in the sector are adept at using big data and data analytics. By analyzing data from diverse sources, companies can identify the best locations for drilling, anticipate when machinery might need maintenance and much more. With the use of advanced analytics, companies turn this raw data into actionable insights in real-time so executives and managers can decide how best to operate their business. The oil and gas industry is now using big data and analytics to achieve higher efficiency levels. It will use the two to customize predictive models and optimize drilling processes. Predictive analytics like variable analysis and condition-based monitoring allow oil and gas companies to design scenario-based simulations and predict future maintenance requirements. That way, they can perform required maintenance before equipment gets damaged.

• *Internet of things* (IoT): This is essentially a network of smart devices that talk to each other. IoT has created a new era of connectivity; devices and machines now communicate easily to create a more efficient and productive environment. In the oil and gas industry, IoT enables the real-time monitoring and data collection of various operations. This connectivity leads to more efficient operations management, better communication in the field, and minimized equipment downtime. More oil and gas companies will use IoT devices to monitor pumps, pipes, and filters to avoid costly leaks. Industrial IoT technology can also help oil and gas companies reduce their overall costs. In modern oil and gas operations, small smart sensors play a big role in maintaining smooth and safe production. These devices can gather data on temperature, pressure, gas levels, and more.

- *Drones:* These are proving to be invaluable in the industry, especially for monitoring oil wells, storage sites, and other facilities. They provide a bird's-eye view that not only speeds up inspections but also enhances the accuracy of assessments. Their simple flight interfaces, flexibility, and ease of adding sensors and accessibility make them a must-have for the oil and gas industry. Drones can safely reach places that are difficult or dangerous for humans, gathering important data without risking lives. Intelligent and flexible tools such as robots and drones reduce human intervention while improving safety and operational efficiency in oil rigs, production facilities, and pipelines. Figure 1.4 shows the use of drones in oil and gas [10].

Figure 1.4 Use of drone in oil and gas [10].

- *Cloud Computing*: This puts data control in the hands of oil and gas companies. They have better analytic insights, more efficient production, and safer drilling. The technology enables companies to scale at speed and accelerate innovation, streamline operations, and become agile. Many oil and gas companies are turning to cloud solutions for data storage, analysis, and management due to their many benefits. The decentralized nature of the cloud makes it much easier for those in the gas and oil industry to perform complex analyses and make decisions. Cloud solutions can assist oil and gas companies in enhancing their sustainability efforts. Large portion of emissions can be avoided by switching from the traditional mode of locally deployed data centers and servers to cloud infrastructure. Cloud computing helps overcome many technical challenges in the oil and gas industry. Cloud-native tools create a functional IT environment online, and we no longer have to invest in hardware or space.

- *Blockchain:* This secures and simplifies complex oil and gas supply chain processes and introduces transparency. Processes such as oil and gas trading, inventory control, shipment tracking, and billing and payments become easier to manage. By logging every deal from parts orders to routes on decentralized, encrypted blocks, blockchain creates a single source of truth. Timestamped records also verify regulatory compliance permanently. This makes it easier for everyone to be on the same page. Blockchain technology increases transparency, accountability, and prevents supply chain disruptions through decentralized payments, cross-chain, and innovative contracts within the oil and gas industry. Figure 1.5 shows the use of blockchain in oil and gas industry [1].

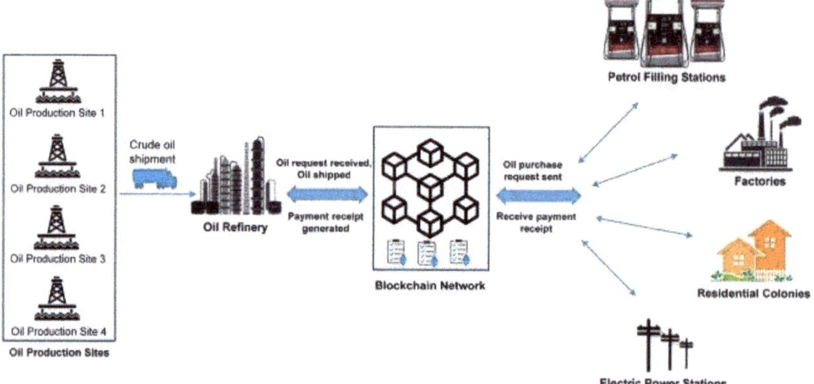

Figure 1.5 Use of blockchain in oil and gas industry [1].

• *Augmented Reality:* AR's popularity in the O&G sector is growing because it can cut down on waste and expenses while improving employee safety. On-site technicians can use AR to instantly connect with off-site specialists, significantly reducing travel costs for flying experts to remote facilities. With this use of AR technology, experts can guide technicians through tasks with audio and video - decreasing the possibility of costly or dangerous errors on-site. AR can also be used to overlay relevant data on the field. Virtual platforms can help employees get up to speed and test their knowledge about what to do in times of crisis.

• *Virtual Reality* (VR): This differs from AR in that it brings users into an entirely digital world, instead of just overlaying images on top of our own reality. Virtual reality enhances training, simulation, visualization, inspections, and remote assistance. Like AR, VR's popularity in enterprise organizations is steadily increasing. Oil and gas companies can cut costs by up to 25 percent per barrel if they use digital solutions like augmented reality and virtual reality. Immersive technology solutions, such as augmented reality (AR), virtual reality (VR), and mixed reality (MR) combine real and virtual environments to increase efficiency and reduce errors. Like digital twins, mixed reality tools merge digital and physical worlds in real-time. They are becoming increasingly useful for companies that manage complex machines and field-based services. Figure 1.6 shows AR, VR, and MR [1].

Figure 1.6 Augmented, virtual, and mixed realities [1].

- *3D Printing:* Additive manufacturing (or 3D printing) facilitates the production of parts on-site to reduce traditional supply chain costs and time massively. This is revolutionizing parts prototyping in the O&G industry and providing cost savings while developing more efficient ways to develop and maintain equipment. For example, Shell used a 3D printed scale model to validate structure design and avoid potential construction issues and cost overruns.

- *Cybersecurity:* Cybersecurity solutions are indispensable for enhancing the safety of upstream, midstream, and downstream oil and gas operations because cybersecurity attacks can result in the loss of confidential information or disruption. It centers around a three-level approach of assessing, protecting, and responding in the event of a cyber threat to improve an O&G company's cyber readiness.

1.5 BENEFITS

By employing a variety of cutting-edge technologies, businesses can benefit from data-led insights, respond in real-time to new challenges, and boost efficiency and productivity. With the right systems and technologies in place, oil and gas industry can tap into their data to gain efficiency, boost resilience, and gain the competitive advantages they need to thrive. Other benefits of emerging technology in the O&G sector include the following [1]:

- *Environmental Concern:* Environmental considerations play an increasingly important role in oil and gas production. Increasing supplies of reliable energy and decreasing emissions is not an either-or proposition. We want to do both, strengthen energy security and help advance the energy transition. For example, advanced sensors do not just boost efficiency, they can also have a significant environmental impact. They provide the visibility required for cutting emissions and can reveal where reductions are possible. As environmental concerns grow, regulatory bodies are tightening their grip on O&G operations, particularly regarding emissions and waste management. Figure 1.7 shows pollution in the oil and gas industry [11].

Figure 1.7 Pollution in oil and gas industry [11].

• *Automation:* One of the most significant advancements in the oil and gas industry is the automation of various processes. Tasks that previously exposed workers to hazardous manual labor can now be automated to limit risk. Through process automation, oil and gas companies can enhance operational efficiency and achieve competitive advantage. Automated drilling systems have revolutionized drilling operations. Automation and robotics are playing pivotal roles in reducing human error, improving safety, and enhancing operational efficiency, particularly in hazardous environments. Automation in the oil and gas industry creates safer work environments by removing manual work processes and accelerating operations. For example, automating drilling processes such as pressure drilling and pipe handling reduce costs and improves safety.

• *Renewable Energy Integration:* To remain competitive, O&G companies must innovate in alternative energy production while continuing to optimize traditional methods. The oil and gas industry is undergoing a significant transformation with the integration of renewable energy sources. The industry is investing in renewable technologies such as solar power, wind energy, and geothermal energy to reduce carbon emissions and improve sustainability. Renewable energy technologies are sufficiently being used to generate electricity for offshore platforms, replacing the need for diesel generators.

• *Price Volatility:* Oil and gas companies face considerable price fluctuations driven by many forces. Digital tools can help firms better predict and respond to this volatility. The key is using data with artificial intelligence to model potential pricing scenarios. This allows companies to forecast supply and demand to make strategic moves.

• *Change Management:* The oil and gas industry constantly shifts as markets, politics, and technology evolve. Digital tools can help companies be ready to adapt quickly.

Predictive analytics track early signals in the wider landscape, like new policies or industry conversations. This heads-up lets leaders avoid big changes by implementing new training programs or adjusting strategies.

• *Predictive Maintenance:* Replacing equipment too early or too late can wreak havoc on your business and budget. With modern technology, clear dashboards can help you recognize deteriorating conditions days or weeks early, allowing targeted repairs before failures. Predictive algorithms with big data also optimize budgets by precisely allocating spending. This predictive maintenance can protect your bottom line from the costs of penalties and incidents. Sustaining consistent output and minimizing downtime also provides more reliability in a traditionally unstable industry.

• *Efficiency:* By leveraging real-time analytics and AI-powered tools, companies can streamline production processes and make informed decisions that enhance operational efficiency. Integrating tech tools to reduce emissions helps a business grow and prosper and be more efficient and competitive in the highly volatile O&G industry.

• *Sustainability:* This is no longer a choice but a necessity. Both consumers and investors are demanding that companies reduce their environmental footprint and adopt greener practices. Sustainability is within reach for the oil and gas industry. And it is clearer than ever that modernizing business models and workflows with disruptive tech is an important step.

• *Expanded Productivity:* Digitization enables easy access to critical industry information right from exploration and production to environmental monitoring and safety. This information helps enhance productivity, reduce risks, and improve all aspects of oil and gas operations.

• *Increased Safety*: Tools like automation and robotics are now an integral part of the drilling process. Connecting and disconnecting drill pipes are perilous operations for humans. Technologies such as iron roughneck robotics and snake-arm robots have completely automated this process, removing all forms of human life risks.

• *Improved Performance*: Gas and oil producers are now harnessing industry tech tools that integrate AI technologies to improve performance, predict trends, and reduce costs. They also use the tools for resource exploration to map and identify petroleum deposits and detect equipment failure and gas leaks.

1.6 CHALLENGES

The oil and gas industry is currently facing many challenges, such as geopolitical implications, unstable crude oil and gas prices, the need to address global climate change, the need to find new sources of oil and gas, and the need to lower emissions. Digital transformation is challenging, costly, resource-intensive, and has lengthy implementation and migration cycles. Other challenges of emerging technology in oil and gas sector include the following [1]:

- *Fears:* Policymakers worry that a significant reduction in US petroleum imports will have geopolitical implications beyond increased US energy security and could alter diplomatic relationships with oil-producing countries. However, these fears are unfounded due to a popular mischaracterization of the role of oil in shaping diplomatic relationships.

- *Old Machinery:* Much of the industry was built on machinery designed for the previous century. While they were reliable, these aging systems lack the connectivity and visibility needed to detect and adapt to today's increasingly dynamic conditions. Outdated monitoring puts safety at risk if threats go unnoticed.

- *Cost:* Investing in new digital technologies can modernize operations, but upgrades often come with big price tags. Leaders must strike the right balance between funding cutting-edge innovations and fixing aging infrastructure. Some companies postpone digital investments, but cutting costs today by delaying technology improvements will eventually backfire.

• *Training:* Employees who are used to old ways of working may need help adopting new software and digital processes. Training programs should fit specific teams like offshore crews, engineers, project managers, and field technicians. You can customize your training to meet the needs of your specific workforce and customers.

• *Regulation:* As oil and gas operations go digital, companies need strategies to follow regulations and protect against hacking. With sites worldwide, companies must follow different rules, depending on their location and the work being done. For example, offshore drilling has many environmental regulations companies must follow. Oil and gas companies are increasingly embracing ESG (environmental, social, and governance) strategies as a way to comply with regulations and appeal to a growing base of environmentally conscious investors.

• *Data Management:* Collecting and managing more data can provide insights to improve efficiency and guide better decisions across operations. However, making the most of this data presents a set of challenges.

- *Complexity:* With today's complex energy assets, no single engineer can design an entire system alone. While technological advancements play a crucial role, other factors such as sustainability initiatives, regulatory pressures, and market demands are also significantly influencing the sector. As companies navigate these complexities, they are finding innovative ways to adapt and thrive in an ever-evolving landscape.

1.7 CONCLUSION

The oil and gas business continues to witness remarkable technological advancements that enhance efficiency, productivity, and sustainability. There is a need for O&G companies to invest in emerging technologies to reduce risks, increase productivity, and reduce costs. Some of these include artificial intelligence, robotics, machine learning, IoT, drones, big data, data analytics, augmented reality, virtual reality, and blockchain. The new technologies in the oil and natural gas sector have enabled the explosion of production growth in the United States. They reinvent how the oil and gas sector operates. As emerging technologies continue to evolve, the oil and gas industry will undoubtedly embrace technologies to address future challenges and contribute to global energy security. The oil and gas industry is still miles behind digital transformation. The gradual adoption of these emerging technologies may lead to new challenges, but over time their adoption will lead to a safer workplace. More information on emerging O&G technologies is available from the books in [12-16].

REFERENCES

[1] A. Dennis, "Digital transformation of the oil & gas industry (2024)," July 2024,

https://whatfix.com/blog/oil-gas-digital-transformation/

[2] "Edge-native AI –Revolutionizing the oil & gas industry,"

https://micro.ai/blog/edge-native-ai-revolutionizing-the-oil-gas-industry

[3] M. N. O. Sadiku, P. A. Adekunte, and J. O. Sadiku, "Emerging technologies in oil and gas industry," *International Journal of Trend in Scientific Research and Development,* vol. 8, no. 5, September-October 2024, pp. 681-688.

[4] M. Halaweh, "Emerging technology: What is it?" *Journal of Technology Management & Innovation*, vol. 8, no. 3, 2013, pp. 108-115.

[5] N. Duggal, "Top 18 new technology trends for 2023," July 2023,

https://www.simplilearn.com/top-technology-trends-and-jobs-article

[6] D. Koroteev and Z. Tekic, "Artificial intelligence in oil and gas upstream: Trends, challenges, and scenarios for the future," *Energy and AI*, vol. 3, March 2001.

[7] "Innovating energy: Emerging technologies in oil and gas production," March 2024,

https://www.dwenergygroup.com/innovating-energy-emerging-technologies-in-oil-and-gas-production/

[8] "Top 8 oil and gas industry trends for 2024," September 2024,

https://kissflow.com/solutions/energy/oil-and-gas-technology-trends/

[9] "Machine learning in the oil and gas industry," November 2023,

https://wezom.com/blog/machine-learning-in-the-oil-and-gas-industry

[10] D. Evans, "4 Emerging technologies disrupting the oil & gas industry in 2019,"

https://irisblog.thewild.com/4-emerging-technologies-disrupting-oil-gas-industry-2019#:~:text=Drones%2C%20augmented%20reality%20(AR),technologies%20influencing%20the%20industry%20today.

[11] "Top 10 trends shaping the oil and gas industry," Unknown Source

[12] J. S. Gomes (ed.), *New Technologies in the Oil and Gas Industry*. IntechOpen, 2012.

[13] Institute for Improved Oil Recovery, Society of Petroleum Engineers (U.S.), *Emerging Technologies for Improved Oil and Gas Recovery: A Technical/economic Assessment: Proceeding.* Institute for Improved Oil Recovery, 1990.

[14] H. Al-Megren and R. Altamimi (eds.), *Advances in Natural Gas Emerging Technologies.* IntechOpen, 2017.

[15] N. K. Mitra, *AI and Digital Technology for Oil and Gas Fields.* Boca Raton, FL: CRC Press, 2024.

[16] D. Dwivedi, A. Ranjan, and J. S. Sangwai (eds.), *Functional Materials for the Oil and Gas Industry (Emerging Trends and Technologies in Petroleum Engineering).*

Boca Raton, FL: CRC Press, 2023.

CHAPTER 2

ARTIFICIAL INTELLIGENCE IN OIL AND GAS

"Artificial intelligence is here and being rapidly commercialized, with new applications being created not just for manufacturing but also for energy, healthcare, and oil and gas. This will change how we all do business."

– Joe Kaeser

2.1 INTRODUCTION

The impact of AI on all spheres of human life is hard to miss. AI has penetrated medicine, education, industry, business, and finance. Machine learning, for example, is a type of AI in which the program analyzes and identifies patterns in the data and optimizes performance without being directed to do so. As a result, the machine learns and improves automatically.

"Data is the new oil" is the highly used phrase these days, and in the case of oil and gas industry, it is a perfect metaphor. Natural gas and oil companies would not be successful at finding oil and gas deposits without technology. Overall, the oil and gas sector (O&G) has long been putting a premium on tradition and caution rather than innovation. Now, all companies in the industry use the latest technology to improve their processes, increase productivity, and protect their market. Many companies are actively working with AI in the oil and gas industry, developing effective solutions. AI may very well be the investment priority for oil and gas companies wanting to see big advances. Through advanced data analysis and machine learning, AI has optimized operations, improved efficiency and safety, enabling more autonomous and predictive drilling.

Artificial intelligence (AI), as the most important general-purpose technology of today, is rapidly entering industries, creating significant potential for innovations and growth. It is reshaping industries everywhere. It has revolutionized various sectors, including the oil and gas industry by applying advanced data analytics and machine learning to increase efficiency and effectiveness in hydrocarbon production. The oil and gas companies play a vital role in the economy. They are the latecomers to digitalization, but they are also getting more and more dependent on AI solutions. The benefits of AI for the oil and gas sector are unparalleled [1].

This chapter discusses in detail and demonstrates how AI is transforming the oil and gas sector. It begins with explaining what artificial intelligence is all about. It describes the oil and gas industry. It covers artificial intelligence in oil and gas. It highlights the benefits and challenges of AI in oil and gas. The last section concludes with comments.

2.2 WHAT IS ARTIFICIAL INTELLIGENCE?

The term "artificial intelligence" (AI) is an umbrella term John McCarthy, a computer scientist, coined in 1955 and defined as "the science and engineering of intelligent machines." It refers to the ability of a computer system to perform human tasks (such as thinking and learning) that usually can only be accomplished using human intelligence [2]. Typically, AI systems demonstrate at least some of the following human behaviors: planning, learning, reasoning, problem-solving, knowledge representation, perception, speech recognition, decision-making, language translation, motion, manipulation, intelligence, and creativity.

The 10 U.S. Code § 2358 defines artificial intelligence as [3]:

1. "Any artificial system that performs tasks under varying and unpredictable circumstances without significant human oversight, or that can learn from experience and improve performance when exposed to data sets.

2. An artificial system developed in computer software, physical hardware, or other context that solves tasks requiring human-like perception, cognition, planning, learning, communication, or physical action.

3. An artificial system designed to think or act like a human, including cognitive architectures and neural networks.

4. A set of techniques, including machine learning, that is designed to approximate a cognitive task.

5. An artificial system designed to act rationally, including an intelligent software agent or embodied robot that achieves goals using perception, planning, reasoning, learning, communicating, decision making, and acting."

AI provides tools for creating intelligent machines that can behave like humans, think like humans, and make decisions like humans. The main goals of artificial intelligence are [4]:

1. Replicate human intelligence

2. Solve knowledge-intensive tasks

3. Make an intelligent connection of perception and action

4. Build a machine that can perform tasks that require human intelligence

5.   Create some system that can exhibit intelligent behavior, learn new things by itself, demonstrate, explain, and can advise to its user.

AI is not a single technology but a range of computational models and algorithms. The concept of AI is an umbrella term that encompasses many different technologies. AI is not a single technology but a collection of techniques that enables computer systems to perform tasks that would otherwise require human intelligence. The major disciplines in AI include [5]:

- *Expert systems*

- *Fuzzy logic*

- *Neural networks*

- *Machine learning (ML)*

- *Deep learning*

- *Natural Language Processors (NLP)*

- *Robots*

These computer-based tools or technologies have been used to achieve AI's goals. Each AI tool has its own advantages. Using a combination of these models, rather than a single model, is recommended. Figure 2.1 shows a typical expert system, while Figure 2.2 illustrates the AI tools. These tools are gaining momentum across every industry. Analytics can be considered a core AI capability. Figure 2.3 shows a concept illustration for the use of modern artificial intelligence technologies for the production of oil and gas [6].

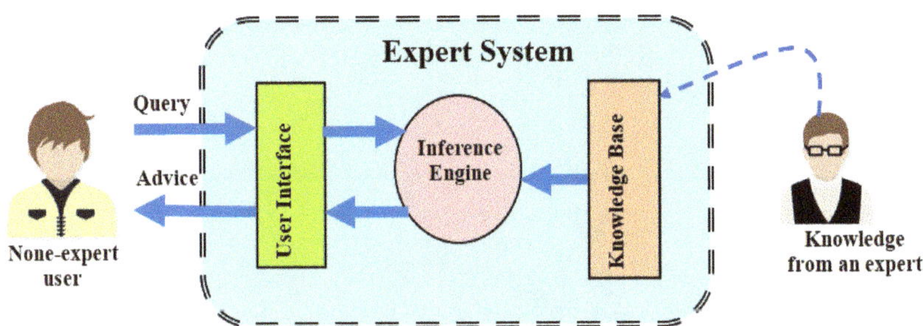

Figure 2.1 A typical expert system.

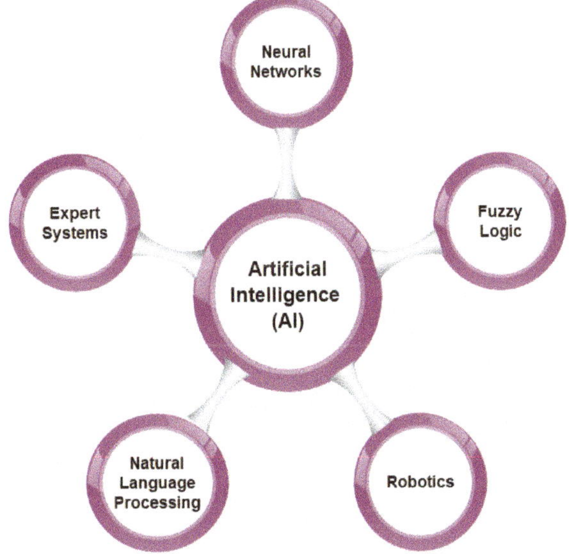

Figure 2.2 AI tools.

Figure 2.3 Concept illustration for the use of AI for production [6].

2.3 OIL AND GAS INDUSTRY

The oil and gas industry is complex and diverse, as typically shown in Figure 2.4 [7]. The industry deals with lots of data coming from manufacturing processes. It divides into upstream, midstream, and downstream, as shown in Figure 2.5 [8]. The upstream summarizes the subsurface (mining) part of the industry, including exploration, field development, and production of crude oil/gas. The upstream segment is the most capital-intensive part of oil and gas and the segment of enormous uncertainties to tackle. It covers crude oil and natural gas production. It includes searching for potential underground or underwater crude oil and natural gas fields, drilling exploratory wells, and subsequently drilling and operating the wells used to lift the crude oil or raw natural gas to the surface. Companies from the sector deal with enormous uncertainties handled manually and relied on expert knowledge, not the actual data.

Figure 2.4 The complexity of the oil and gas industry [7].

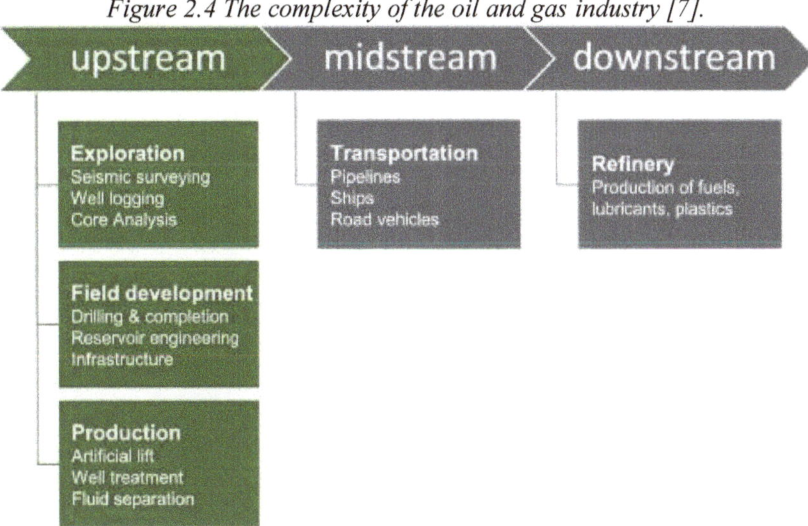

Figure 2.5 The oil and gas industry divides into upstream, midstream, and downstream [8].

Midstream stands for storage and transportation of oil and gas. In midstream business, AI can support proper planning and execution, optimal route selection, etc. In contrast, it helps refiners plan optimal blending, forecasting the demand, estimating prices, and improvising customer relationships in the downstream business.

Downstream is for refinery and distribution, i.e., production of fuels, lubricants, plastics, and other products. Figure 2.6 shows a downstream operation [9]. By embracing AI, downstream firms achieve refinery and distribution cost reduction and reach regulatory compliance faster in several ways. Beyond crude oil and natural gas, downstream oil and gas companies benefit from AI-driven predictions of consumer demand for various products. The application of AI in downstream oil and gas companies involves real-time monitoring systems that optimize refineries. Those systems monitor operations and collect data during distillation, catalytic cracking, and hydrogenation. They also screen data from energy meters and equipment sensors. Downstream operations can be optimized to minimize costs and maximize spreads. AI predictive maintenance reduces AI downstream processes and optimizes maintenance planning.

Figure 2.6 A downstream operation [9].

2.4 ARTIFICIAL INTELLIGENCE IN OIL AND GAS

The oil and gas industry is highly competitive, and companies are always looking for ways to increase efficiency and reduce costs. Like every industry today, the O&G sector is trying to figure out how to adopt and deploy artificial intelligence and its tools. These forms of AI are used in the oil and gas industry [9]:

- *Machine learning:* Machine learning (ML) is a specific area that has seen significant deployment in oil and gas in recent decades. It analyzes data for pattern recognition to predict outcomes. ML is often used in reservoir exploration, drilling, or fault detection. Machine learning for oil and gas reduces the risk of drill-bit failures and optimizes extraction rates.

• *Deep learning:* As a more advanced form of ML, deep learning utilizes complex neural networks for tasks like precise seismic analysis to process data and identify more complex details within it.

• *Generative AI:* This is a sub-category of artificial intelligence (AI) that aims to help the user create an image, video, or song using different instructions called user "prompts." It learns from existing datasets, gen AI creates, for example, new data samples, emergency instructions, or smart summaries. Using gen AI, oil and gas companies and logistics providers automate mission-critical supply chain processes. Generative AI in oil and gas adds resilience to planning so all stakeholders make the supply chain run like clockwork. In the oil and gas industry, Generative AI is a groundbreaking tool that can transform how oil and gas experts tackle tasks, interpret data, and convey intricate concepts.

• *Natural Language Processing (NLP) and Computer Vision:* Interpret human language and visual data for tasks like report generation and quality control.

- *Edge AI*: Processes data locally on IoT devices without relying on cloud storage or Internet connectivity.

- *Intelligent Robots:* Robots designed with AI capabilities for hydrocarbon exploration and production, to improve productivity and cost-effectiveness while reducing worker risk.

- *Virtual Assistants:* Online chat platform that helps customers navigate product databases and processes general inquiries using natural language.

Figure 2.7 shows AI use cases in oil and gas sector [10].

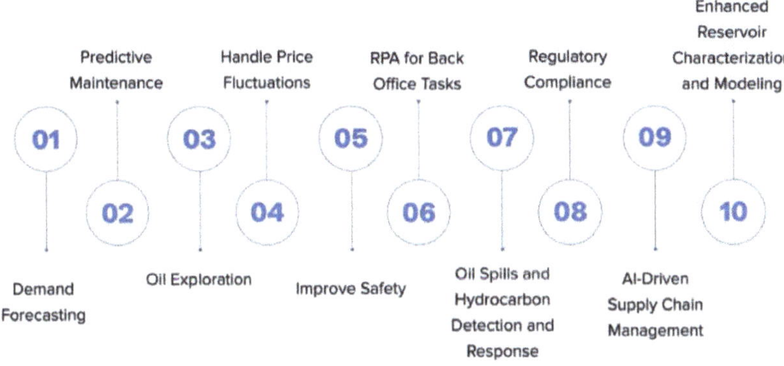

Figure 2.7 AI use cases in oil and gas sector [10].

2.5 BENEFITS

AI in the oil and gas industry brings numerous benefits, including enhanced efficiency, cost reduction, and improved safety. Artificial intelligence technology is being put to use in certain highly specific applications. By processing and analyzing the big data, AI algorithms are now aiding us in solving complex problems. When utilized together, AI and big data solutions have an almost limitless range of potential applications and can optimize a wide variety of operations within the energy sector to deliver improved efficiency, reliability, and sustainability. AI helps oil and gas companies assess the value of specific reservoirs, customize drilling, and completion plans according to the geology of the area, and assess risks of each individual well. Other benefits of AI in O&G section include the following [11].

- *Safety*: The most prevalent problem in coal mining is safety. The oil and gas industry needs to increase the need to improve safety and reduce environmental impact. The risk of injury in oil and gas plants and mines is extremely high. Failure to comply with safety rules and protocols can lead to irreversible damage to health and even death. The industry is inherently risky, and accidents can have severe consequences for both workers and the environment. AI technology can help companies improve safety by identifying potential hazards and implementing measures to mitigate them. No one has cause to worry that machine learning technology will soon replace manual labor entirely. Instead, it will make skilled workers more efficient, spare them unnecessary tasks and, most importantly, improve workplace safety.

•   *Automation:* By automating routine tasks and optimizing complex operations, AI enables companies to streamline their processes and reduce operational costs. Employees' routine tasks are being automated by AI technology, which will have the impact of freeing up time for employees to contribute more that is directly relevant to their field of expertise. AI has contributed to the development of autonomous drilling systems, which can automatically analyze and adjust drilling parameters in real-time to increase efficiency and safety.

•   *Monitoring:*   AI computer   vision solutions can monitor the workplace. They can alert management even about the smallest of regulatory deviations. The only purpose is to keep workers safe from serious harm.

•   *Predictive Intelligence:* With the proliferation of sensors and other data-gathering technologies, companies have access to more data than ever before. Companies are heavily investing in AI-powered solutions that can analyze data from sensors and other sources to improve efficiency and minimize downtime. AI technology can help companies make sense of this data and extract insights that can inform decision-making and drive operational improvements. Predictive intelligence allows engineers to convert cross-sourced historical and real-time data into actionable insights for drilling preparation.

• *Predictive Maintenance:* AI-powered predictive maintenance models have revolutionized operations in the oil and gas industry. It enables the implementation of condition-based maintenance methods for critical infrastructure. Using machine learning techniques, AI can predict failures and perform predictive maintenance on key equipment and machinery. This helps reduce operating costs and minimize unplanned downtime.

• *Operations Optimization:* AI can analyze vast amounts of operational data, such as pressure, temperature, and flow measurements, to identify hidden patterns and trends. This allows companies to streamline their operations and improve production efficiency.

• *Image and Video Analysis:* AI enables automated analysis of images and videos captured in the oil field, making it easier to detect anomalies, leaks, and damage, thereby improving risk management and safety.

• *Optimizing Hydrocarbon Recovery*: AI can improve reserve estimation accuracy and help optimize production strategies, leading to greater hydrocarbon recovery and more efficient exploitation of reservoirs.

• *Emissions Reduction:* By improving efficiency in energy production and consumption, AI can help reduce greenhouse gas emissions associated with the oil and gas industry.

• *Risk management:* AI can analyze real-time and historical data to assess risk and assist in strategic decision-making to protect infrastructure and assets in risky situations.

• *Cost Reduction:* The oil industry faces formidable challenges to reduce costs, including volatile oil prices, escalating operational expenses, and the need for substantial investments in new technologies. Work in upstream oil and gas industries was once considered highly labor-intensive, but AI has helped change this perspective. Companies are now maximizing production while minimizing costs, all thanks to AI-powered process automation and predictive maintenance.

• *Greater Decision-Making*: In mining, fast and accurate decision-making is crucial. AI is the bridge between decision-makers and accurate data trends. Operators can use AI to make better exploration and production decisions and optimize acquisition strategies with better forecasts of lease transaction prices. Critical decisions can now be made much faster and more accurately.

- *Innovation and Improvement:* AI technologies, including machine learning and predictive analytics, are revolutionizing traditional practices in oil and gas. By automating data analysis and operational processes, AI enables companies to innovate in areas such as exploration, drilling, and production.

- *Better Customer Experience*: The oil industry faces significant challenges in delivering exceptional customer experiences. AI enhances customer interactions by improving service delivery and responsiveness. Oil companies can utilize AI to predict customer needs and streamline supply chain operations, ensuring timely delivery of products.

- *Competitive Advantage:* The adoption of AI provides a significant competitive edge by enabling companies to operate more efficiently and make data-driven decisions. AI helps in optimizing resource allocation, predicting market trends, and enhancing operational agility, which is crucial for staying ahead in a highly competitive market.

- *Environmental Protection:* AI plays a crucial role in promoting sustainability within the oil and gas sector by optimizing processes to reduce emissions and improve energy efficiency, AI helps companies adhere to environmental regulations. Many oil and gas companies are actively seeking to reduce their carbon footprint and make their operations greener, and AI comes to the forefront in these endeavors.

Some of these benefits or advantages are summarized in Figure 2.8 [12].

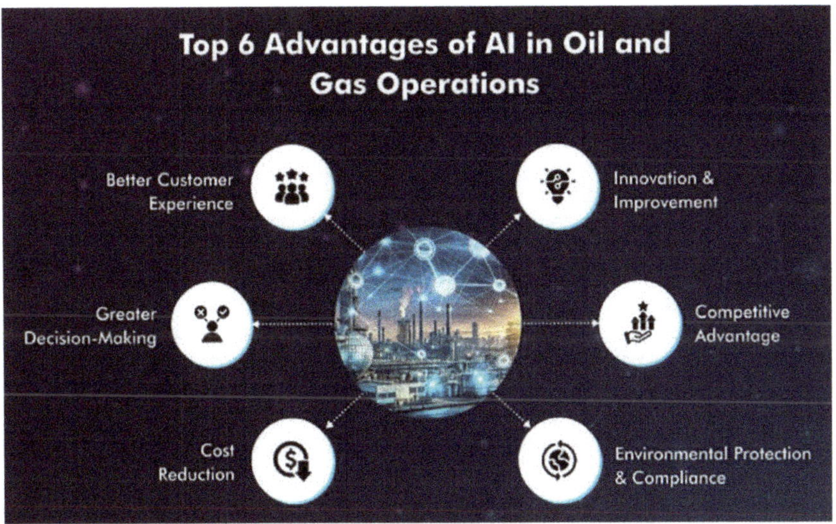

Figure 2.8 Some of the advantages AI in O&G operations [12].

2.6 CHALLENGES

While some oil and gas companies, like BP, Shell, Saudi Aramco, and Gazprom Neft are jump-starting their AI initiatives by investing aggressively in startups and R&D, several challenges are preventing them from massively and rapidly implementing AI in the exploration and production of oil and gas. AI initiatives will not fail because of bad algorithms, but rather because of lack of vision, late or even no changes in the organization's operational and business model, due to lack of high-resolution data and poor collaboration. Other challenges of AI in O&G section include the following [8]:

- *Experts:* The success of artificial intelligence critically depends on human intelligence. To actively use AI in processes and products, companies must grow in-house teams composed of data and AI specialists. This means that oil and gas companies will become (partially) data-driven companies and, that AI specialists will become irreplaceable in supporting almost all innovation efforts in oil and gas companies in the future.

- *Shortage of Talents*: Finding and retaining AI talent is a challenging task. There is a significant shortage of AI talent on the job market, and with more and more companies getting into AI and forming their own AI groups, prospects are not good for the next decade. This is especially true for oil and gas companies. Hundreds of thousands of Baby Boomers heading into retirement at the same time when fewer and fewer young people have chosen to major in areas like petroleum engineering, geology, and other key areas of expertise relevant to the industry.

- *Partnership*: Next to working more with data and data scientists, petroleum engineers will have to learn how to work with AI assistants – products similar to Alexa and Siri, but focused on industry applications. In these new partnerships, the challenge will be to combine the best from the two sides. AI is expected to be dominantly used by humans to augment their decision-making abilities rather than replace them.

- *Quality Data*: AI tools need good quality data of a suitable volume to be trained and then to work properly in the operational mode. Access to big and quality data is a crucial enabler and barrier for AI applications' successful development. Oil and gas fields generate large amounts of raw data. Still, it is not a guaranty for success as there are known issues with the quality and accuracy of field data and overall lack of large volumes of labeled data in the oil and gas industry. To enhance the value of data oil and gas companies possess or can access, they will have to redesign and adjust their organizational structures and processes.

- *Collaboration*: Artificial intelligence is born in open and collaborative environment as a consequence of academia being a leading force in AI research for decades, almost without any business influences. This created a culture of free sharing. Oil and gas companies are not famous for their joint industry projects, especially between competitors and especially not in strategic domains such as AI. They have to embrace open collaboration as a standard to succeed in the era of AI once they join the race. The need to acquire the latest AI technology and talent are additional reasons for oil and gas companies to adopt open collaboration.

• *Cybersecurity:* The integration of AI systems increases vulnerability to cyber threats, as interconnected systems can expose sensitive operational data. The rise in the number of physical and cyberattacks and its security spending have necessitated the need for artificial intelligence tools to encrypt the working system into their enterprise's security. Video cameras as sensors help in monitoring the security threats in the utilities all day.

• *Cultural Resistance*: The oil and gas industry is traditionally conservative, leading to resistance among employees and stakeholders towards adopting AI technologies. This cultural inertia can hinder innovation and the effective implementation of AI solutions. Employees may be reluctant to embrace new technologies, fearing job displacement or unfamiliarity with AI-driven processes.

Customer preferences and expectations may also resist changes to established practices.

- *Regulatory Compliance:* Navigating the complex regulatory landscape is a major challenge for AI deployment in the oil and gas sector. Companies must ensure compliance with various regulations concerning data privacy, environmental standards, and safety protocols. With the global energy demand escalating, companies in the oil and gas field face growing demands to boost operational efficiency, cut expenses, and adhere to safety and environmental regulations.

Some of these challenges are summarized in Figure 2. 9 [12].

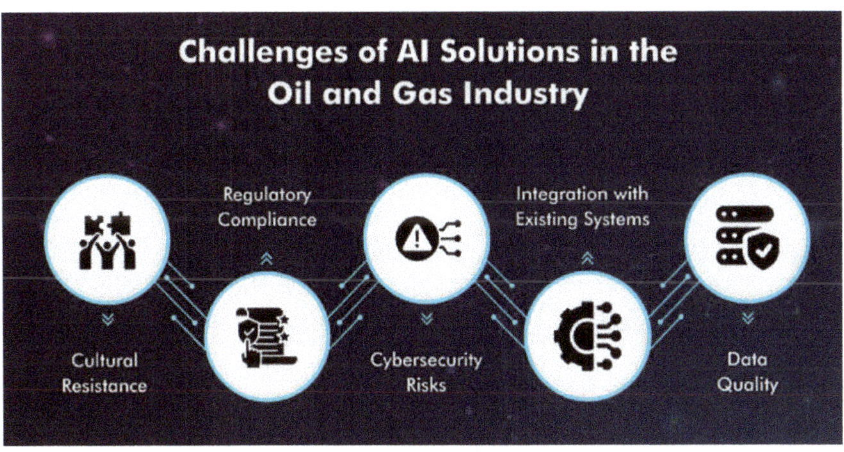

Figure 2.9 Some of the challenges of AI in O&G industry [12].

2.7 CONCLUSION

AI is an exciting and diverse scientific field. If deployed ethically and properly, it has the potential to positively impact every field of endeavor. The integration of artificial intelligence into the oil and gas industry represents a pivotal shift towards addressing the complex challenges of today's energy landscape. The O&G industry has been increasingly adopting AI technology in recent years, and this trend is expected to continue. It recognizes the power and impact that artificial intelligence can have on the industry's performance. The Oil and Gas Authority (OGA) is deploying AI in many ways, which will assist them in discovering new oil & gas projections and enhancing production from existing infrastructures. More information on artificial intelligence in O&G industry is available from the books in [13-20] and the following related journals:

- *Energy and AI*

- *The AI Journal*

REFERENCES

[1] M. N. O. Sadiku, P. A. Adekunte, and J. O. Sadiku, "Artificial intelligence in oil and gas industry," *International Journal of Trend in Scientific Research and Development*, vol. 8, no. 5, September-October 2024, pp. 689-697.

[2] M. N. O. Sadiku, "Artificial intelligence," *IEEE Potentials,* May 1989, pp. 35-39.

[3] "Artificial intelligence (AI),"

https://www.law.cornell.edu/wex/artificial_intelligence_(ai)

[4] "Artificial intelligence tutorial,"

https://www.javatpoint.com/artificial-intelligence-tutorial

[5] D. Quinby, "Artificial intelligence and the future of travel," May 2017,

https://www.phocuswright.com/Travel-Research/Research-Updates/2017/Artificial-Intelligence-and-the-Future-of-Travel

[6] D. Blackmon, "How will AI be applied to oil and gas? One company has some answers," May 2024,

https://www.forbes.com/sites/davidblackmon/2024/05/07/how-will-ai-be-applied-to-oil--gas-one-company-has-some-answers/

[7] "AI and big data,"

https://www.aramco.com/en/what-we-do/energy-innovation/digitalization/ai-and-big-data?utm_source=&utm_medium=&utm_campaign=&utm_term=&utm_content=&gad_source=1&gclid=EAIaIQobChMI88bijvH1iAMVwEh_AB02ei8OEAAYASAAEgL8t_D_BwE

[8] D. Koroteev and Z. Tekic, "Artificial intelligence in oil and gas upstream: Trends, challenges, and scenarios for the future," *Energy and AI*, vol. 3, March 2001.

[9] "Hype aside: Real-world use cases of artificial intelligence in the oil and gas industry,"

https://medium.com/instinctools/hype-aside-real-world-use-cases-of-artificial-intelligence-in-the-oil-and-gas-industry-a8c6f12fd10d

[10] "Unleashing the potential of artificial intelligence in the oil and gas industry – 10 use cases, benefits, examples," September 2024,

https://appinventiv.com/blog/artificial-intelligence-in-oil-and-gas-industry/

[11] O. Ostos, "Influence of artificial intelligence on oil and gas production systems,"

https://www.linkedin.com/pulse/influence-artificial-intelligence-oil-gas-production-systems-ostos

[12] "AI for oil industry: Key challenges and innovative solutions," August 2024,

https://medium.com/predict/ai-for-oil-industry-key-challenges-and-innovative-solutions-5ddce257a89d#:~:text=The%20oil%20industry%20faces%20signi ficant,ensuring%20timely%20delivery%20of%20products.

[13] M. N. O. Sadiku, S. M. Musa, and S. R. Nelatury, *Applications of Artificial Intelligence.* Sherida, NY: Gotham Books, 2022.

[14] M. Ahmadi, *Artificial Intelligence for a More Sustainable Oil and Gas Industry and the Energy Transition: Case Studies and Code Examples.* Elsevier Science, 2024.

[15] C. Mohan, *AI and GenAI in Oil and Gas Industry: Revolutionizing Oil and Gas With Artificial Intelligence.* Kindle Edition, 2024.

[16] Y. N. Pandey et al., *Machine Learning in the Oil and Gas Industry: Including Geosciences, Reservoir Engineering, and Production Engineering with Python.* Apress, 2020.

[17] P. Bangert, *Machine Learning and Data Science in the Oil and Gas Industry: Best Practices, Tools, and Case Studies.* Gulf Professional Publishing, 2021.

[18] P. R. Chellah et al., *The Power of Artificial Intelligence for the Next-Generation Oil and Gas Industry: Envisaging AI-inspired Intelligent Energy Systems and Environments.* Wiley-IEEE Press, 2023.

[19] M. Shah, A. Kshirsagar, and J. Panchal, *Applications of Artificial Intelligence (AI) and Machine Learning (ML) in the Petroleum Industry.* Boca Raton, FL: CRC Press, 2022.

[20] A. Hemmati-Sarapardeh et al., *Applications of Artificial Intelligence Techniques in the Petroleum Industry*. Gulf Professional Publishing, 2020.

CHAPTER 3

BIG DATA IN OIL AND GAS

"The goal is to turn data into information, and information into insight."

– Carly Fiorina

3.1 INTRODUCTION

The world has seen a digital revolution where more and more work is being conducted online. Storing this activity has led to the concept of big data, i.e., large datasets that are otherwise difficult to manage. Because of these characteristics, big data requires new technologies and techniques to capture, store, and analyze. The cloud word for big data is shown in Figure 3.1 [1]. Typical sources of big data are shown in Figure 3.2 [2]. Big data is of great interest to oil and gas production and operation. With the advent of new technologies, there has been a massive increase in the amount of data generated within oil and gas sector.

Figure 3.1 The cloud word for big data [1].

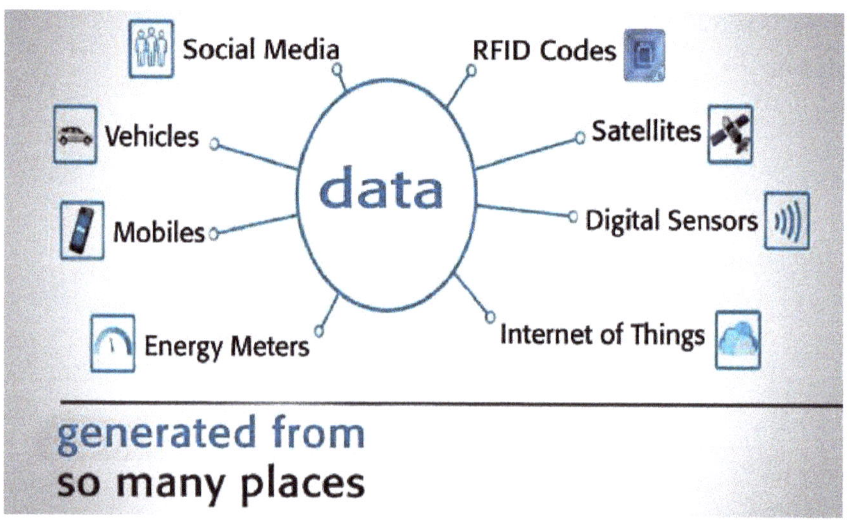

Figure 3.2 Typical sources of big data [2].

The oil and gas sector is one of the largest and most complex industries in the world, involving the exploration, extraction, refining, and distribution of hydrocarbon resources. The complexity of the O&G operation is illustrated in Figure 3.3 [3]. With the recent advent of data recording sensors in exploration, drilling and production operations, oil and gas industry has become a massive data-intensive industry. There are ample opportunities for oil and gas companies to use big data to get more oil and gas out of hydrocarbon reservoirs, reduce capital and operational expenses, increase the speed and accuracy of investment decisions, and improve health and safety while mitigating environmental risks. Big data can be used to improve decision-making and operational efficiency by analyzing the data to uncover patterns and correlations. Other new technologies such as deep learning, cognitive computing, and augmented and virtual reality can be used to extract useful information enormously reducing the data processing time.

Figure 3.3 The complexity of the O&G operation [3].

With the recent introduction of data recording sensors in exploration, drilling, and production processes, the oil and gas industry has transformed into a massively data-intensive industry. These data can come from sensors, data recording devices, spatial and GPS coordinates, weather services, and seismic data. Since the data recording devices and sensors are different in types, the generated data can be in different sizes and formats. The vast quantity of data is challenging to be handled due to storage, sustainability, and analysis issues. The main application of big data is to provide processing and analysis tools for increasing amounts of data. Big data analyzes huge data sets to reveal the underlying trends and help engineers forecast potential issues [4].

This chapter reviews the utilization of big data and data analytics in the oil and gas industry. It begins with explaining what big data is all about and mentions its characteristics. It describes the oil and gas industry and big data in the industry. It covers big data in oil and gas. It presents some applications of big data in oil and gas. It highlights the benefits and challenges of big data in oil and gas. The last section concludes with comments.

3.2 WHAT IS BIG DATA?

Big data applies to data sets of extreme size (e.g. exabytes, zettabytes) which are beyond the capability of the commonly used software tools. It involves situations where very large data sets are big in volume, velocity, veracity, and variability [5]. The data is too big, too fast, or does not fit the regular database architecture. It may require different strategies and tools for profiling, measurement, assessment, and processing.

Big Data is essentially classified into three types [6]:

- *Structured Data:* This is highly organized and is the easiest to work with. Any data that can be stored, accessed, and processed in the form of fixed format is known as structured data. It may be stored in tabular format. Due to their nature, it is easy for programs to sort through and collect data. Structured data has quantitative data such as age, contact, address, billing, expenses, credit card numbers, etc. Data that is stored in a relational database management system is an example of structured data.

- *Unstructured Data:* This refers to unorganized data such as video files, log files, audio files, and image files. Any data with an unknown form or structure is classified as unstructured data. Almost everything generated by a computer is unstructured data. It takes a lot of time and effort is required to make unstructured data readable. Examples of unstructured data include Metadata, Twitter tweets, and other social media posts.

- *Semi-structured Data:* This falls somewhere between structured data and unstructured data, i.e., both forms of data are present. Semi-structured data can be inherited such as location, time, email address, or device ID stamp.

The different types of big data are depicted in Figure 3.4 [7].

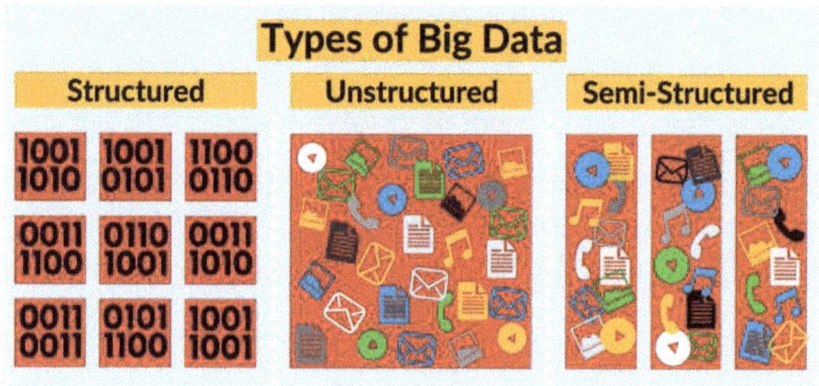

Figure 3.4 Types of big data [7].

The process of examining big data is often referred to as big data analytics. It is an emerging field since massive computing capabilities have been made available by e-infrastructures [8]. Analytics include statistical models and other methods that are aimed at creating empirical predictions. Data-driven organizations use analytics to guide decisions at all levels. Several techniques have been proposed for analyzing big data. These include the HACE theorem, cloud computing, Hadoop, and MapReduce [9].

3.3 CHARACTERISTICS OF BIG DATA

Big data is growing rapidly and expanding in all science and engineering, including physical, biological, and medical services.

Different companies use different means to maintain their big data. As shown in Figure 3.5 [10], big data is characterized by 42 Vs. The first five Vs are volume, velocity, variety, veracity, and value [2].

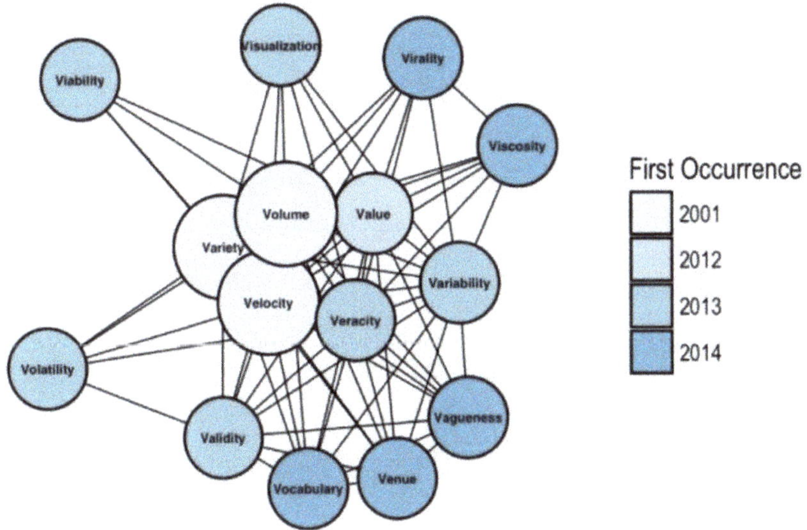

Figure 3.5 The 42 V's of big data [10].

- *Volume*: This refers to the size of the data being generated both inside and outside organizations and is increasing annually. Some regard big data as data over one petabyte in volume.

- *Velocity*: This depicts the unprecedented speed at which data are generated by Internet users, mobile users, social media, etc. Data are generated and processed in a fast way to extract useful, relevant information. Big data could be analyzed in real-time, and it has movement and velocity.

- *Variety*: This refers to the data types since big data may originate from heterogeneous sources and is in different formats (e.g., videos, images, audio, text, logs). BD comprises of structured, semi-structured or unstructured data.

- *Veracity*: By this, we mean the truthfulness of data, i.e. whether the data comes from a reputable, trustworthy, authentic, and accountable source. It suggests the inconsistency in the quality of different sources of big data. The data may not be 100% correct.

- *Value*: This is the most important aspect of the big data. It is the desired outcome of big data processing. It refers to the process of discovering hidden values from large datasets. It denotes the value derived from the analysis of the existing data. If one cannot extract some business value from the data, there is no use managing and storing it.

On this basis, small data can be regarded as having low volume, low velocity, low variety, low veracity, and low value. Additional five Vs has been added [11]:

- *Validity:* This refers to the accuracy and correctness of data. It also indicates how up-to-date it is.

- *Viability:* This identifies the relevancy of data for each use case. Relevancy of data is required to maintain the desired and accurate outcome through analytical and predictive measures.

- *Volatility:* Since data are generated and change at a rapid rate, volatility determines how quickly data change.

- *Vulnerability:* The vulnerability of data is essential because privacy and security are of utmost importance for personal data.

- *Visualization:* Data needs to be presented unambiguously and attractively to the user. Proper visualization of large and complex clinical reports helps in finding valuable insights.

Instead of the 10V's above, some suggest the following 5V's: Venue, Variability, Vocabulary, Vagueness, and Validity) [12].

Industries that benefit from big data include the healthcare, financial, airline, travel, restaurants, automobile, sports, agriculture, and hospitality industries. Big data technologies are playing an essential role in farming: machines are equipped with sensors that measure data in their environment. Structured and unstructured data are generated in various types [13-15].

3.4 OIL AND GAS INDUSTRY

Oil is difficult to locate. The oil reservoirs are typically found 5,000 to 35,000 feet below the earth's surface, making them hard to find.

Oil is an expensive commodity, and a lot of science, engineering, and workforce are required to produce oil. Given the cost, quantity, and availability of oil, the companies involved in this industry must identify methods to stay profitable. Big data analytics has benefited the oil and gas industry in many ways.

The oil industry is divided into upstream, midstream, and downstream, as shown in Figure 3.6 [16] and explained as follows:

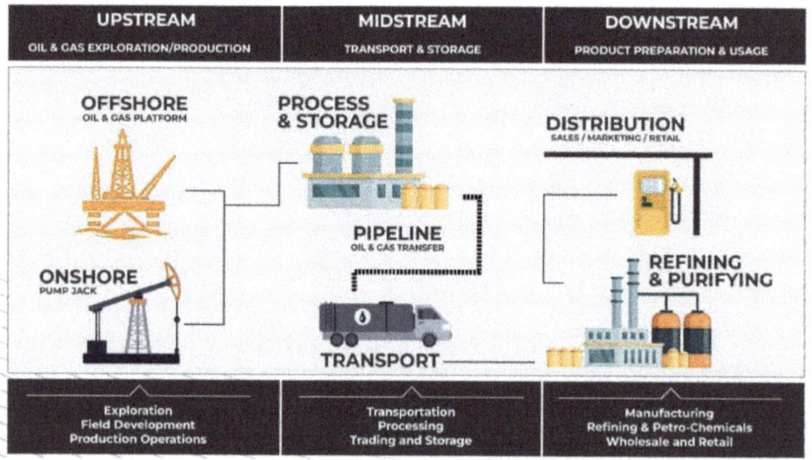

Figure 3.6 The oil industry divides into upstream, midstream, and downstream [16].

- *Upstream Sector:* The upstream process in oil and gas operations refers to the discovery and production of oil and gas. Many activities are done in the upstream area, wherein big data analytics plays a crucial role. Among all business segments upstream segment is the most dominant segment owing to the increasing use of big data analytics for the discovery of non-conventional shale gas. Upstream analytics begins with the collection of seismic data with sensors across a potential area of interest looking for petroleum sources. Then the data is aggregated, cleaned, processed, and analyzed to choose the best location for drilling. Figure 3.7 shows the upstream operation [17].

Figure 3.7 Upstream operation [17].

- *Midstream Sector:* The midstream activities in the oil and gas industry refer mainly to the transportation of oil and gas, i.e., logistics. Big data analytics is used to enhance shipping performance. For example, to improve the performance of ships and reduce greenhouse emissions, big data analytics helps by predicting the propulsion power. Big data analytics is essential for planning pipelines and infrastructure to transport oil from sources to refineries and pumping stations. Its significance in logistics makes it a critical tool for oil and gas companies, given the highly flammable nature of the transported material.

- *Downstream Sector:* The downstream is responsible for refining petroleum products and delivering them to end-users. It mainly involves refining and selling oil and gas. It is expected to be the second largest segment due to the increasing use of product analytics solutions which assist refineries in the standard chemical composition of the finished products. The downstream oil and gas industry is undergoing a significant transformation due to the integration of big data technologies. The sector is leveraging big data in numerous ways to enhance efficiency and sustainability.

Plenty of raw information is available for analysis in the oil and gas industry. Whether you are involved in upstream, midstream, downstream, administration, or the commodities market, you are surrounded by data.

3.5 BIG DATA IN OIL AND GAS

Big data in oil and gas (O&G) is one of the prominent technologies that are now disrupting the industry with innovative methods. It gives businesses and business owners a huge potential to move forward and grow. More industry executives believe that big data is the solution to boost their business operations. For example, major O&G companies like ExxonMobil, BP, and Shell heavily invest in big data and AI solutions. The companies that embrace and implement new technologies are positioning themselves as frontrunners in shaping the industry's future landscape. To make high profits, these companies must make an environment-friendly commitment and apply IoT for price and asset monitoring.

3.6 APPLICATIONS OF BIG DATA IN OIL AND GAS

The oil and gas industry is no stranger to data. For decades, it has generated massive volumes of information from exploration, production, and distribution operations. The recent technological improvements have resulted in daily generation of massive datasets in oil and gas exploration and production industries. Big data in the oil and gas industry is the massive amount of data generated by various processes and transactions in the sector. The advent of modern sensing technologies, such as the Internet of things (IoT), and remote satellite monitoring, has created an explosion in data generation throughout the sector. Figure 3.8 shows data analytics in oil and gas sector [16], while Figure 3.9 shows the key components of big data analytics in oil and gas industry [18]. Here are some ways big data is used in the oil and gas industry [19,20]:

Figure 3.8 Data analytics in oil and gas sector [16].

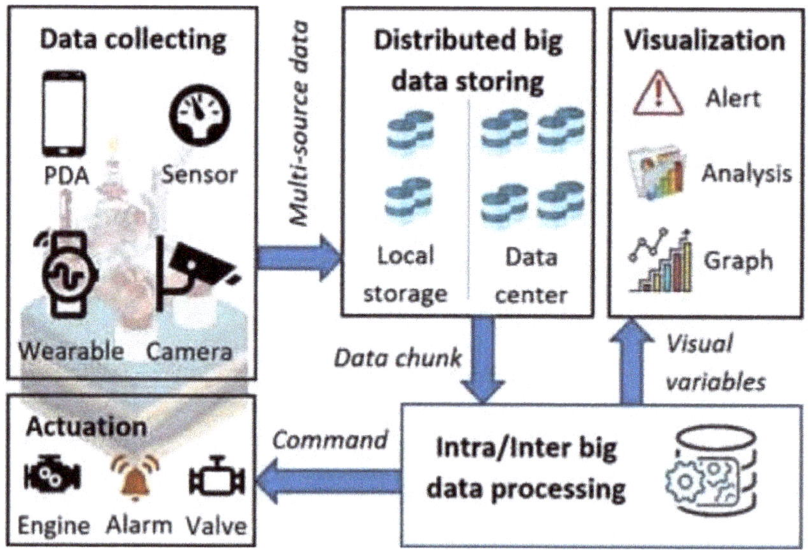

Figure 3.9 Key components of big data analytics in oil and gas industry [18].

- *Supply Chain Optimization:* One key area of application is supply chain optimization. By optimizing logistics, tracking inventory levels, and improving distribution efficiency, downstream companies reduce operational costs while minimizing waste--thus ensuring that products reach customers in a timely manner. Big data can help optimize supply chains by analyzing data on inventory, transportation, demand, and market trends. Data-driven demand forecasting models help companies manage their inventories, optimize logistics, and make informed decisions for both upstream and downstream operations.

• *Assessing New Prospects:* Competitive intelligence is created using analytics applied to geospatial data, oil and gas reports and other syndicated feeds in order to bid for new prospects.

• *Enhanced Oil Recovery:* Enhancing oil recovery from existing wells is a key objective for oil and gas companies. Analytics applied to a variety of big data at once (seismic, drilling, and production data) could help reservoir engineers map changes in the reservoir over time and provide decision support to production engineers for making changes in lifting methods. This type of approach could also be used to guide fracking in shale gas plays.

• *Real-time Production Optimization:* Real-time SCADA and process control systems combined with analytics tools help O&G producers optimize resource allocation and prices by using scalable compute technologies to determine optimum commodity pricing. They also, help to make more real-time decisions with fewer engineers. Big data can be used to monitor and analyze production processes in real-time, which can help identify bottlenecks and predict equipment failures.

• *Prevent Cyber-Terror Acts:* Oil companies need to identify events or patterns that could indicate an impending security threat or cyber-terrorist act in order to keep their personnel, property and equipment safe. Predictive analytics is the central part of identifying patterns that can help detect these threats beforehand.

• *Data Crunching:* From wells, digs, and extraction to transportation and refining, there are ample opportunities to collect data on every single aspect of your business operations. Investing in machine learning will help a company absorb petabytes of sensor data from drills faster than a whole fleet of workers on the actual job site. This frees up the people to make better, more informed decisions.

• *Predictive Analysis:* Algorithms can use historical data and experience to make better and more enlightened predictions about future operations. Companies in possession of this level of enhanced analysis can use it to make very educated guesses about future trends, prices, production, and actions in the market. The more you understand what could happen with your industry, the better prepared you will be for any contingency.

• *Decision-Making:* Data plays a critical role in the decisions that create value. Today, data analytics is being leveraged throughout the oil and gas value chain to optimize decision-making and improve overall performance. Operators make decisions every day in the field, typically with limited involvement by central functions or subject matter experts. With predictive analysis, you can make faster decisions, supported by faster delivery of decision support information, to identify possible threats. You can make the move from educated guesses to confirmed, real-time decisions with ease and confidence. In essence, data analytics does not remove the human element from making decisions. It helps humans make better decisions.

• *Predictive Maintenance:* Big Data analytics enable predictive maintenance through IoT devices and sensors deployed in downstream facilities. Predictive maintenance models based on data analysis can proactively identify areas requiring maintenance and reduce equipment failures, leading to optimized operational efficiency and reduced costs. Everyday operations in the oil and gas industry depend upon a substantial amount of machinery. Thus, it is essential for a business to keep tabs on the condition and fitness of its equipment so it can address problems before a shutdown occurs. Advanced analytics can compare the age of a given machine with its rate of past and future usage to determine when it is most likely to need maintenance and replacement.

- *Data Mining:* This is the extraction of relevant information and insights from large datasets using statistical and computational methods. Data mining is an integral part of big data analytics, which entails processing, analyzing, and interpreting large and complex datasets to discover patterns, trends, and insights that can assist organizations in making informed decisions. Data cleansing, data validation, data normalization, and data transformation are some of the methods used by data mining and analytics practitioners to reduce the likelihood of these errors.

3.7 BENEFITS

Big data is used to identify conditions or anomalies that would impact drilling operations. Big data solutions can provide companies in the oil and gas industry with insights into exploration, drilling, and production processes to ensure their optimization, reduce environmental risks, streamline equipment maintenance planning, enhance oil recovery, and more. Big data also has other benefits for the oil and gas industry, including the following [21]:

- *Safety:* The responsibility for the health and safety of the individuals working in the oil and gas industry is with the companies that employ them. Big data can be used to monitor for safety hazards and environmental risks, such as leaks, air quality, and pipeline integrity. Big data analytics helps prevent accidents by predicting and detecting anomalies and issues, such as stress corrosion and fatigue cracks in pipes and trucks, along with early detection of seismic movements.

- *Cost-Effectiveness:* The effectiveness of big data analytics in the oil and gas industry has been widely acknowledged as it has evolved into a crucial instrument for enhancing operational efficiency and lowering costs. When you have access to game-changing information about your operations, it becomes easier to increase the efficiency and efficacy of your projects. This increase in productivity can be experienced at every level of the oil and gas industry. This streamlined supply chain not only enhances cost-effectiveness but also contributes to a more sustainable operation by reducing the environmental footprint associated with transportation and storage.

- *Manage Seismic Data:* Drilling for oil in deep water can cost over $100 million, and it is crucial that you find the right location. In fact, to avoid any risks and save time and money, Shell uses fiber optic cables and the data is then transferred to its private servers.

- *Optimize Drilling Process*: Today's oil drilling platforms have about 80,000 sensors, which are expected to generate 15 petabytes of data during the lifetime of a platform. The large amount of data gathered by these sensors allows for predictive maintenance of equipment and timely replacements in order to reduce downtime and increase efficiency.

- *Improve Reservoir Engineering:* Big data solutions help oil and gas companies to collect, process, and analyze data that is essential for making reservoir production more effective. They help to collect and process data that O&G companies need to make reservoir production more effective. This data is collected using a number of downhole sensors (temperature sensors, acoustic sensors, pressure sensors, and others). To make energy more affordable and sustainable, we use big data tools to understand the earth's subsurface better.

- *Improve Logistics:* The oil and gas industry faces the problem of safely transporting petroleum. The major problem that concerns logistics in the oil and gas industry is transporting petroleum while reducing risks as much as possible. To ensure that gas and oil are transported safely, companies use sensors and predictive maintenance. Sensors, predictive maintenance, and other technologies help detect any faults in pipelines or tankers. This allows for safe logistics of petroleum products.

• *Quality:* Just as with petroleum data analytics, the use of big data analytics can be instrumental in maintaining consistent product quality. By monitoring and controlling the refining process in real-time, companies can ensure that their products meet stringent specifications and safety standards. This proactive approach also enhances customer trust and brand reputation.

• *Workforce Management*: Data-driven insights help employers optimize their workforce strategy and develop training programs tailored to specific job roles and skill sets.

• *Prescriptive Analytics:* By incorporating human expertise and external data sources into AI models, companies can generate detailed recommendations for specific actions and their potential outcomes.

• *Data Integration:* The oil and gas industry has focused on data integration, i.e., how do we get all the data in one place and make it available to the geoscientists and engineers working to find and produce hydrocarbons. Proper data integration grows more crucial as oil fields mature because operators must understand changing field conditions. Managers know more than anyone that they must maximize hydrocarbon production while reducing drilling costs.

3.8 CHALLENGES

The sheer volume, complexity, and speed of data generation in the O&G industry have created both significant challenges and opportunities. Leveraging big data in the oil and gas sector poses challenges, such as data integration from diverse sources, data quality assurance, data privacy, and security concerns, and the need for a skilled team to understand and analyze the data. Implementing big data in the oil and gas industry presents challenges related to data transfer, collection frequency, and data quality. They include [21,22]:

- *Legacy Systems:* Many oil and gas firms still rely on legacy systems and outdated technologies, which can hinder the integration and effective utilization of modern data analytics platforms.

- *Historical Data:* The oil and gas industry faces the challenge of handling vast amounts of historical data accumulated over decades. This historical data, often in various formats and from different sources, needs to be integrated and made compatible with modern operations.

Accumulated and analyzed historical data of various injury-causing accidents help identify patterns and trends to mitigate the risk of working in this field. Overcoming this challenge requires robust data management strategies and investments in data infrastructure to facilitate seamless data transfer.

• *Data Integration:* Merging data from disparate sources and systems can be a complex and resource-intensive task. The establishment of a unified data model (or data lake) is crucial for successful analytics implementation.

• *Data Quality:* The data available in the oil and gas industry is unique and peculiar. Data quality and accuracy issues are among the serious problems that are present in the industry. Ensuring data accuracy, reliability, and timeliness is critical for analytics success, otherwise, the insights drawn might be misleading or even detrimental. Stringent data governance mechanisms and maintaining data integrity should be prioritized.

• *Scalability:* Oil and gas companies should have scalable analytics infrastructure and agile analytics platforms that can evolve and handle massive amounts of data while adapting to the ever-changing industry landscape.

- *Talent and Skills:* The oil and gas industry has traditionally been an engineering-driven domain, but success in the big-data era requires diverse talents and expertise to navigate the multi-disciplinary landscape of analytics. In addition to that, you need a thorough understanding of the physics of the problem.

- *High Cost*: One of the major challenges of big data's application in any industry including oil and gas industry is the cost associated with managing the data recording, storage, and analysis. There is a high financial cost associated with dealing with data. This includes various data management activities such as data recording, storage, maintenance, and analysis. A huge investment is involved in the infrastructure of energy pipelines; hence its integrity is a must for reliable operations.

3.9 CONCLUSION

Big data is the idea that some aspect of your business operations generates large amounts of information, and you need to figure out what to do with it. No matter how entrenched in the "old ways" a company might be, now is the time for the entire oil and gas industry to embrace the benefits of big data analytics. Since O&G companies create lots of information, it makes sense to find new and improved ways to put that information to the best possible use.

The integration of advanced analytics and the increasing reliance on big data are driving the oil and gas sector toward a future where fully autonomous control systems for complex processing facilities become a reality. The future of big data in the oil and gas industry is promising, with the potential to drive efficiency, safety, and sustainability. However, realizing these benefits requires careful planning, investment, and a commitment to addressing challenges such as data security and integration. More information about big data in O&G operations can be found in the books in [23-31] and the following related journals:

- *Petroleum*

- *Energy Reports*

- *Oil & Gas Journal*

REFERENCES

[1] R. Delgado, "The challenges of bringing BYOD to the military,"

https://socpub.com/articles/the-challenges-of-bringing-byod-to-the-military-11272

[2] J. Moorthy et al., "Big data: Prospects and challenges," *The Journal for Decision Makers*, vol. 40, no. 1, 2015, pp. 74–96.

[3] "Repsol launches big data, AI project at Tarragona refinery," June 2018,

https://www.ogj.com/refining-processing/refining/operations/article/17296578/repsol-launches-big-data-ai-project-at-tarragona-refinery

[4] M. N. O. Sadiku, C. M. M. Kotteti, and J. O. Sadiku, "Big data in oil & gas industry," *International Journal of Science, Engineering and Technology*, vol. 12, no. 5, 2024, pp. 1-10.

[5] M. N.O. Sadiku, M. Tembely, and S.M. Musa, "Big data: An introduction for engineers," *Journal of Scientific and Engineering Research,* vol. 3, no. 2, 2016, pp. 106-108.

[6] "The complete overview of big data,"

https://intellipaat.com/blog/tutorial/hadoop-tutorial/big-data-overview/

[7] R. Allen, "Types of big data | Understanding & Interacting with key types (2024),"

https://investguiding-com.custommapposter.com/article/types-of-big-data-understanding-amp-interacting-with-key-types

[8] P. Baumann et al., "Big data analytics for earth sciences: the earthserver approach," *International Journal of Digital Earth*, vol. 19, no. 1, 2016, pp.3-29.

[9] X. Wu et al., "Knowledge engineering with big data," *IEEE Intelligent Systems*, September/October 2015, pp.46-55.

[10] "The 42 V's of big data and data science,"

https://www.kdnuggets.com/2017/04/42-vs-big-data-data-science.html

[11] P. K. D. Pramanik, S. Pal, and M. Mukhopadhyay, "Healthcare big data: A comprehensive overview," in N. Bouchemal (ed.), *Intelligent Systems for Healthcare Management and Delivery.* IGI Global, chapter 4, 2019, pp. 72-100.

[12] J. Moorthy et al., "Big data: Prospects and challenges," *The Journal for Decision Makers*, vol. 40, no. 1, 2015, pp. 74–96.

https://www.grandviewresearch.com/industry-analysis/industrial-wireless-sensor-networks-iwsn-market

[13] A. K. Tiwari, H. Chaudhary, and S. Yadav, "A review on big data and its security,"

Proceedings of IEEE Sponsored 2ⁿᵈ International Conference on Innovations in Information Embedded and Communication Systems, 2015.

[14] M. B. Hoy, "Big data: An introduction for librarians," *Medical Reference Services Quarterly*, vol. 33, no 3. 2014, pp. 320-326.

[15] M. Viceconti, P. Hunter, and R. Hose, "Big data, big knowledge: Big data for personalized healthcare," *IEEE Journal of Medical and Health Informatics*, vol. 19, no. 4, July 2015, pp. 1209-1215.

[16] "Oil-gas industry and big data analytics: How data analytics is impacting oil & gas industry," February 2024,

https://www.analytixlabs.co.in/blog/data-analytics-in-oil-and-gas/

[17] "Hype aside: Real-world use cases of artificial intelligence in the oil and gas industry,"

https://medium.com/instinctools/hype-aside-real-world-use-cases-of-artificial-intelligence-in-the-oil-and-gas-industry-a8c6f12fd10d

[18] S. Srivastava, "Big data analytics in the oil and gas industry – Benefits, use cases, examples, challenges," September 2024,

https://appinventiv.com/blog/big-data-analytics-in-oil-and-gas/

[19] E. Brancaccio, "Big data in oil and gas industry,"

https://www.oil-gasportal.com/big-data-in-oil-and-gas-industry/?print=pdf

[20] "Using big data analytics for oil & gas,"

https://eaginc.com/big-data-analytics-oil-gas-industry/

[21] "Benefit from big data analytics in the oil and gas industry," September 2023,

https://wezom.com/blog/benefit-from-big-data-analytics-in-the-oil-and-gas-industry

[22] M. Jensen, "The big data boom - How data analytics is revolutionizing the oil & gas industry," January 2024,

https://www.linkedin.com/pulse/big-data-boom-how-analytics-revolutionizing-oil-gas-matthew-jensen-vuipc

[23] M. N. O. Sadiku, U. C. Chukwu, and P. O. Adebo, *Big Data and Its Applications.* Moldova, Europe: Lambert Academic Publishing, 2024.

[24] K. R. Holdaway, *Harness Oil and Gas Big Data with Analytics: Optimize Exploration and Production with Data Driven Models.* John Wiley & Sons, 2014.

[25] J. Gohil and M. Shah, *Application of Big Data in Petroleum Streams.* Boca Raton, FL: CRC Press, 2022.

[26] K. Srivastava et al. (eds.), *Understanding Data Analytics and Predictive Modelling in the Oil and Gas Industry.* Boca Raton, FL: CRC Press, 2023.

[27] K. R. Holdaway and D. H. B. Irving, *Enhance Oil and Gas Exploration with Data-Driven Geophysical and Petrophysical Models.* Wiley, 2017.

[28] A. S. Al-Harrasieh, *Exploring the Factors Impacting the Adoption of Big Data Analytics: Oil and Gas Industry in Oman.* Sultan Qaboos University, 2012.

[29] A. Baaziz, *How to Use Big Data Technologies to Optimize Operations in Upstream Petroleum Industry*. SSRN, 2015.

[30] F. Aminzadeh, L. A. Zadeh, and M. Nikravesh (eds.), *Soft Computing and Intelligent Data Analysis in Oil Exploration. Volume 51*. Elsevier Science, 2003.

[31] S. Mishra, and A. Datta-Gupta, *Applied Statistical Modeling and Data Analytics: A Practical Guide for the Petroleum Geosciences*. Elsevier Science, 2017.

CHAPTER 4

INTERNET OF THINGS
IN OIL AND GAS

"If you think that the Internet has changed your life, think again. The Internet of things is about to change it all over again!"

– Brendan O'Brien

4.1 INTRODUCTION

Today, the Internet has become an indispensable part of life. When it comes to the Internet, the Internet of things (IoT) has taken center stage. The IoT is a giant network of connected things and people. The idea behind creating IoT was the amalgamation of the physical world into computer-based systems. Some of the examples of IoT devices are cell phones, washing machines, laptops, etc. With IoT, anything that can be connected shall be connected. This is best illustrated in Figure 4.1 [1]. Internet of things is a tool for connecting physical objects to the virtual world using sensors and some Internet protocols to lessen human interventions. The role of IoT in the oil and gas industry is illustrated in Figure 4.2 [2].

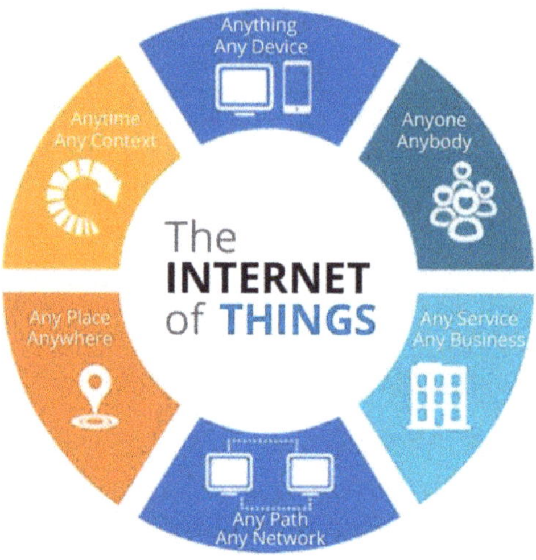

Figure 4.1Anything that can be connected is connected to IoT [1].

Figure 4.2 The role of IoT in the oil and gas industry [2].

The oil and gas (O&G) industry serves as a lifeline for the global economy, powering key sectors such as transportation, electricity generation, heating, and manufacturing. The industry is known for its complex infrastructure, as is typically illustrated in Figure 4.3 [3]. The industry keeps on being progressively focused and companies cannot afford to be left behind. In oil and gas exploration and production, IoT sensors are deployed on drilling equipment and pipelines to monitor parameters such as pressure, temperature, and flow rates. This data helps optimize operations, reduce downtime, and enhance safety. Inexpensive IoT acoustic sensors in the oil field continuously analyze oil composition (oil, water, gas, etc.) within pipelines. Sensors that monitor inventory levels of onshore oil tanks automatically dispatch trucks when the tanks need to be emptied. Sensors also monitor the performance of above-ground pumps to alert maintenance teams of potential and actual issues. Individual IoT sensors connected by fiber optic cables aid oil exploration by mapping subsurface drilling sites to determine new drilling locations and optimize output of operational sites. Figure 4.4 shows seven ways in which IoT is revolutionizing the O&G industry [4].

Figure 4.3 The O&G industry is known for its complex infrastructure [3].

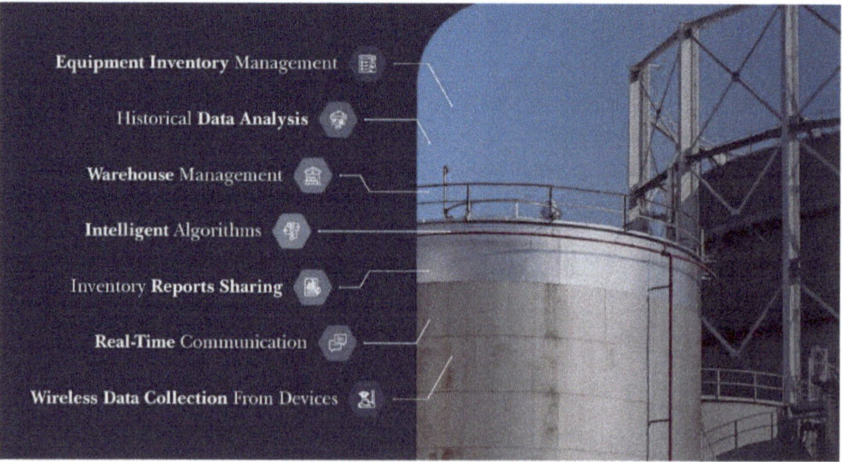

Figure 4.4 Seven ways in which IoT is revolutionizing the O&G industry [4].

The Internet of things (IoT) is a concept that refers to the interconnectedness of everyday objects and devices through the Internet, allowing them to collect, exchange, and analyze data. It is a digital technology that is transforming the oil and gas industry by improving safety, efficiency, and operational performance. It is revolutionizing the world and how we live our lives. IoT devices can range from sensors to industrial machinery, all of which communicate with each other to enhance automation, efficiency, and decision-making. This rapid advancement enables the oil and gas sector to outpace other capital-intensive industries in terms of innovation and efficiency [5].

The objective of this chapter is to provide an informative guide on how IoT is revolutionizing oil and gas exploration. It begins with explaining the concept of Internet of things and also industrial IoT. It describes the oil and gas industry. It presents some applications of IoT in oil and gas. It highlights the benefits and challenges of IoT in oil and gas. The last section concludes with comments.

4.2 CONCEPT OF INTERNET OF THINGS

The term "Internet of things" was introduced by Kevin Ashton from the United Kingdom in 1999. Internet of things (IoT) is a network of connecting devices embedded with sensors. It is a collection of identifiable things with the ability to communicate over wired or wireless networks. The devices or things can be connected to the Internet through three main technology components: physical devices and sensors (connected things), connection and infrastructure, and analytics and applications.

The IoT is a worldwide network that connects devices to the Internet and to each other using wireless technology. IoT is expanding rapidly and it has been estimated that 50 billion devices will be connected to the Internet by 2020. These include smartphones, tablets, desktop computers, autonomous vehicles, refrigerators, toasters, thermostats, cameras, pet monitors, alarm systems, home appliances, insulin pumps, industrial machines, intelligent wheelchairs, wireless sensors, mobile robots, etc. A typical IoT is shown in Figure 4.5 [6].

The Internet of things is much more than a simple technology. There are four main technologies that enable IoT [7]:

(1)   Radio-frequency identification (RFID) and near-field communication.

(2)   Optical tags and quick response codes: This is used for low-cost tagging.

(3)   Bluetooth low energy (BLE).

(4)   Wireless sensor networks: They are usually connected as wireless sensor networks to monitor physical properties in specific environments.

Other related technologies are cloud computing, machine learning, and big data.

The Internet of things (IoT) technology enables people and objects to interact with each other. It is employed in many areas such as smart transportation, smart cities, smart energy, emergency services, healthcare, data security, industrial control, logistics, retails, government, traffic congestion, manufacturing, industry, security, agriculture, environment, and waste management. Figure 4.6 shows the most widely used application areas of IoT [8].

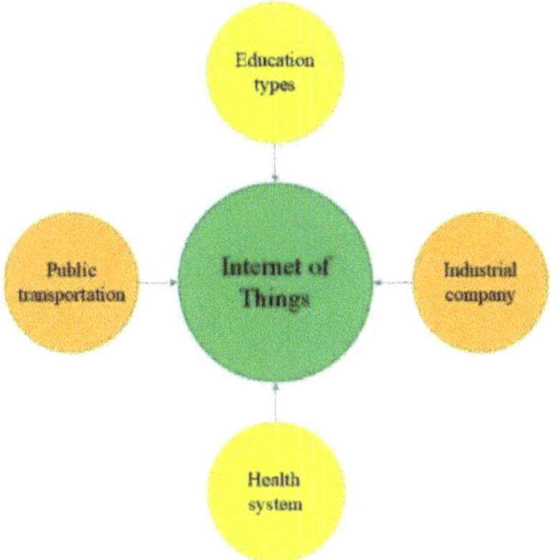

Figure 4.6 The most widely used IoT application areas [8].

IoT supports many input-output devices such as cameras, microphones, keyboards, speakers, displays, microcontrollers, and transceivers. It is the most promising trend in the healthcare industry. This rapidly proliferating collection of Internet-connected devices, including wearables, implants, skin sensors, smart scales, smart bandages, and home monitoring tools has the potential to connect patients and their providers in a unique way.

Today, smartphone acts as the main driver of IoT. The smartphone is provided with healthcare applications.

The narrowband version of IoT is known as narrowband IoT (NBIoT). This is an attractive technology for many sectors including healthcare because it has been standardized [8]. The main feature of NBIoT is that it can be easily deployed within the current cellular infrastructure with a software upgrade.

4.3 INDUSTRIAL INTERNET OF THINGS

The Internet of things (IoT) is essentially the connection of devices to the Internet. IoT network connects various types of devices like tablets, smartphones, personal computers, laptops, and wearable devices. The implementation of IoT in oil and gas sector helps streamline the processing and distribution of oil and gas supplies, boost operational effectiveness, reduce costs, and enhance customer satisfaction with safe and swift deliveries. The industrial Internet of things (IIoT) involves the use of connected devices, advanced software, and sensor-based tools in sync with industrial machinery. There is a strong and growing market for industrial IoT (IIoT) applications in the oil and gas sector. IIoT sensing networks can monitor potential dangers such as gas leaks or oil spills, enabling them to respond immediately with corrective action when something unusual occurs.

4.4 IOT IN OIL AND GAS INDUSTRY

The Internet of things (IoT) in the oil & gas industry is the network of physical objects connected to the Internet. IoT for the oil and gas industry involves implementing interconnected devices, sensors, and systems. It is used to collect, transfer, and analyze raw data in real-time to get a clear picture of processes at the facilities, improve operational efficiency, reduce energy consumption, and drive profitability. IoT solutions cover the three segments of the oil and gas industry [2]:

- *Upstream Sector:* Upstream companies (e.g., exploration and production) focused on optimization can gain new operational insights by analyzing diverse sets of physics, non-physics, and cross-disciplinary data. Companies in the upstream area lose billions of dollars in discovering other onshore or offshore oil and natural gas fields and work O&G repositories can be at a depth of more than 3,000 meters in the seas. That is the reason the utilization of robots and sensors to break down surface and underground situations could spare a huge number of dollars and an enormous measure of time. In addition, the upstream part alone utilizes a large number of specialists who are occupied with a variety of processes, and each function has potential dangers. Actually, oil and gas workers are seven times more likely to be injured. IoT sensors are deployed on drilling rigs, equipment, and wellheads to continuously monitor their condition. IoT devices collect vast amounts of data, which can be analyzed to optimize drilling techniques and predict equipment failures. Implementing the Internet of things in upstream oil and gas can help with more efficient, better management of drilling and extraction processes, reduce unproductive time, and improve overall operational efficiency in the upstream sector.

• *Midstream Sector*: Midstream companies (e.g., transportation, such as pipelines and storage) eyeing higher network integrity and new commercial opportunities will tend to find significant benefits by building a data-enabled infrastructure. These companies face a huge issue of oil and gas leakage. Unexpected pipeline breaks are a genuine migraine for midstream companies all over the globe. The US Department of Transportation estimated that around nine million gallons of crude oil has spilled from pipelines in the US since 2010. Consequently, automated pipeline investigation is the need of great importance. Some companies have introduced sensors inside and outside of the pipelines to feel, see, hear, and smell different parts of their oil pipelines and to recognize potential pipeline breaches. More companies are moving toward building up a data-enabled monitoring infrastructure and employing IoT in the oil and gas industry. IoT sensors are placed along pipelines to monitor factors like pressure, temperature, and flow rates. This data helps detect leaks, corrosion, or other issues promptly. Figure 4.7 shows the midstream operation [10].

Figure 4.7 The midstream operation [10].

• *Downstream Sector:* Downstream players (e.g., petroleum products refiners and retailers) should see the most promising opportunities in revenue generation by expanding their visibility into the hydrocarbon supply chain and targeting digital consumers through new forms of connected marketing. The maturing infrastructure and surprising downtime keep on plaguing the downstream side of the oil and gas industry. In a universe of low oil costs, diminishing spontaneous downtime can be the establishment of operational productivity. If the pump needs substitution early, it gets replaced and a sudden shutdown is avoided. With the utilization of IoT in oil and gas, refineries can improve their execution, limit their downtime, and production keeps on moving consistently. IoT sensors in refineries monitor equipment performance, enabling operators to fine-tune processes for maximum efficiency and reduced energy consumption. IoT can be used to monitor the transportation and distribution of oil and gas products, allowing for real-time tracking and route optimization.

4.5 APPLICATIONS OF IOT IN OIL AND GAS

The IoT in the oil and gas industry is like having eyes everywhere, with physical objects fitted with sensors, software, and network connectivity. These devices communicate with each other in real-time, allowing for seamless task execution. IoT applications in the oil and gas industry empower businesses to automate processes, increase productivity, and decrease costs by collecting and analyzing data to identify areas for improvement. IoT devices can be used for a variety of purposes in O&G operations, including [9]:

- *Safety:* IoT-enabled cameras and GPS devices can identify potential hazards, while sensors can help keep employees safe by reducing the need for manual data collection. IoT applications in oil and gas industry can improve worker and environmental safety. IoT can enhance worker safety through wearable devices that monitor vital signs and environmental conditions. Continuous monitoring of emissions, gas leaks, and other potential environmental hazards using IoT sensors can help prevent accidents and reduce the environmental impact of operations.

• *Monitoring:* IoT devices can monitor equipment performance and identify failures. IoT enables real-time monitoring of drilling operations from a central control room, allowing for immediate response to any issues or deviations from the plan. Sensors can monitor tank levels and detect harmful gasses in the surrounding area. IoT sensors can monitor the condition and performance of critical equipment such as pumps, compressors, and pipelines. This enables predictive maintenance, reducing downtime and optimizing asset utilization. IoT can track environmental conditions near pipelines and storage facilities, helping operators respond quickly to any environmental concerns. Figure 4.8 shows production monitoring [11], while Figure 4.9 shows monitoring of operations [12].

Figure 4.8 Production monitoring [11].

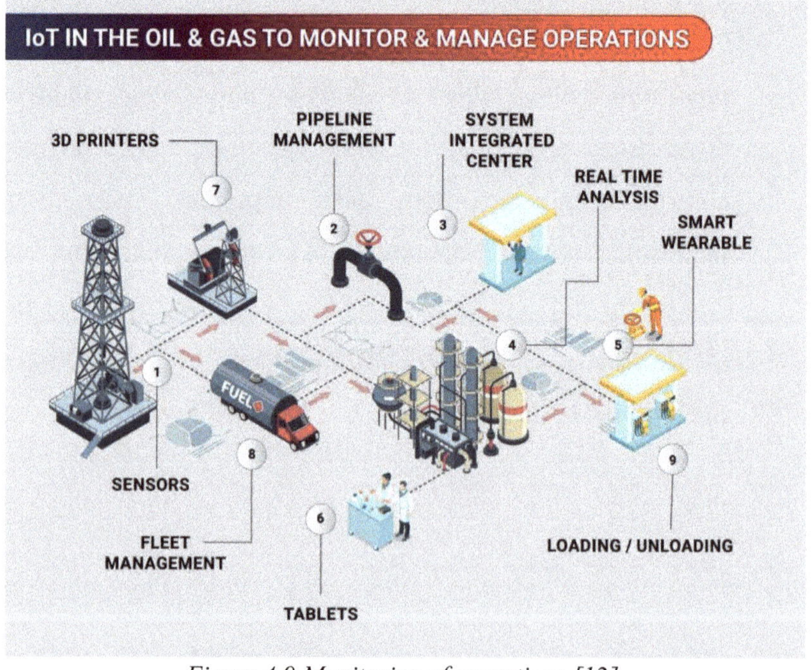

Figure 4.9 Monitoring of operations [12].

- *Supply Chain Efficiency*: IoT can monitor the transportation and distribution of oil and gas products. IoT can provide real-time visibility into the supply chain, including tracking the movement of raw materials, products, and equipment, leading to better logistics and inventory management.

- *Data Collection and Analytics:* This is crucial for decision-making in the oil and gas industry. In all segments, data collected by IoT devices is typically sent to centralized platforms for analysis. The use of IoT in oil and gas enables better data collection, real-time monitoring, and data analytics, allowing companies to make informed decisions, improve operational efficiency, enhance safety, and address environmental concerns. IoT generates vast amounts of real-time data from sensors and devices. This data can be collected and analyzed to gain insights into operations, equipment performance, and environmental conditions. The IoT data collected can be used for predictive maintenance, helping to identify equipment issues before they lead to costly breakdowns.

- *Exploration:* The oil and gas sector has quickly recognized the transformative potential of IoT technology in exploration and production processes. Forward-thinking companies are shifting their attention from merely using IoT devices such as sensors to devising bold strategies for leveraging the data gathered from these devices.

- *Drilling:* Drilling represents a substantial part of oil and gas production costs. Integrating IoT devices boosts operational effectiveness while optimizing water, chemicals, and sand use. IoT sensors can map subsurface drilling sites to help determine new drilling locations and optimize output. IoT devices facilitate the tracking of drilling variables, such as pressure, temperature, and equipment performance, allowing for instant adjustments and enhanced drilling efficiency. Data generated by IoT enables anticipatory maintenance, minimizing downtime, and extending the life of drilling equipment.

- *Predictive Maintenance:* IoT plays a crucial role in predictive maintenance in the oil and gas industry. Predictive maintenance involves using data from IoT sensors and analytics to predict when equipment is likely to fail so that maintenance can be scheduled proactively. It reduces downtime and avoids costly emergency repairs, leading to significant cost savings in terms of maintenance and lost production. IoT sensors, combined with advanced analytics and machine learning algorithms, can predict equipment failures and optimize maintenance schedules. By continuously monitoring the condition of pumps, compressors, and other machinery, oil and gas companies can reduce downtime, extend the life of assets, and lower operational costs.

4.6 BENEFITS

The Internet of things (IoT) has the potential to revolutionize the oil and gas industry by providing real-time data insights, improving operational efficiency, and reducing costs. IoT is revolutionizing the industry by enabling real-time monitoring, predictive maintenance, and data-driven insights. The ability to transfer data without requiring human interaction enables previously unprecedented amounts of data to be collected and exchanged with other devices, or through a central platform. Other benefits of IoT applications in the oil and gas industry include the following [9]:

- *Real-time Monitoring:* The oil and gas industry faces increasing pressure to minimize its environmental impact. IoT sensors can be used to monitor emissions, water quality, and soil conditions in real-time. This data can assist companies in complying with environmental regulations and reducing their carbon footprint.

- *Environmental Monitoring*: IoT-enabled sensors and devices in the oil and gas industry are facilitating real-time monitoring and data collection, enabling proactive maintenance and reducing downtime. IoT technology enables real-time monitoring of equipment and processes. This is achieved through the deployment of various sensors and devices throughout the production and distribution network. IoT devices are equipped with a variety of sensors such as pressure, temperature, humidity, and flow sensors. These sensors continuously collect data from equipment and processes.

• *Energy Management:* IoT can help optimize energy consumption by monitoring and controlling equipment like generators, HVAC systems, and lighting to reduce operational costs.

• *Safety:* IoT enhances safety measures and hazard detection in the oil and gas industry. IoT sensors can detect the presence of harmful gases and volatile compounds in real-time. This information helps operators respond quickly to gas leaks and prevent accidents.

• *Security*: IoT can enhance security by monitoring unauthorized access to critical infrastructure and alerting authorities to potential threats. As IoT adoption increases, so does the vulnerability to cyberattacks. Upcoming IoT technologies will include more advanced cybersecurity measures, such as AI-driven threat detection and encryption protocols, to safeguard critical infrastructure and data.

- *Cost Efficiency:* IoT contributes to cost efficiency in the oil and gas sector. Real-time monitoring ensures that operations are in accordance with regulatory standards, helping companies avoid fines and penalties. IoT technology helps optimize energy consumption by monitoring and controlling equipment remotely, reducing energy waste and operational costs.

- *Energy Efficiency:* Monitoring and simulating energy consumption and production processes can help oil and gas companies identify opportunities for energy efficiency improvements. IoT can play a vital role in optimizing energy consumption within the industry. Smart grids and energy management systems can use IoT data to reduce energy waste and enhance the overall efficiency of operations, resulting in cost savings and reduced environmental impact.

- *Better Decision-Making:* The data collected by IoT devices offer valuable insights into operations, allowing for informed decision-making and strategic planning based on real-time information. Whether you want to optimize processes, cut costs strategically, or identify new business opportunities, IoT-generated insights enable better decision-making with real-time insights.

- *Sustainability*: IoT technologies are crucial in reducing the oil and gas industry's carbon footprint by detecting leaks, improving energy efficiency, and ensuring environmental compliance.

- *Remote Services:* IoT technologies work to make repairs and services safer and more cost-effective in the oil and gas industry. The information collected from each system allows workers to actively monitor overall system performance so they know when to schedule repairs and maintenance. Connected systems can be programmed to send alerts to other connected devices if they begin to fail. Alerts also can be sent when systems are about to experience a high-pressure situation, malfunction, or any other dangerous issues.

- *Automation:* At a time when well drilling and completion complexities are increasing and field experts are becoming scarcer, automation offers many benefits. Delegating work to computers eases a lot of these problems and streamlines the process to make it effective and efficient.

Some of these benefits/advantages are shown in Figure 4.10 [12].

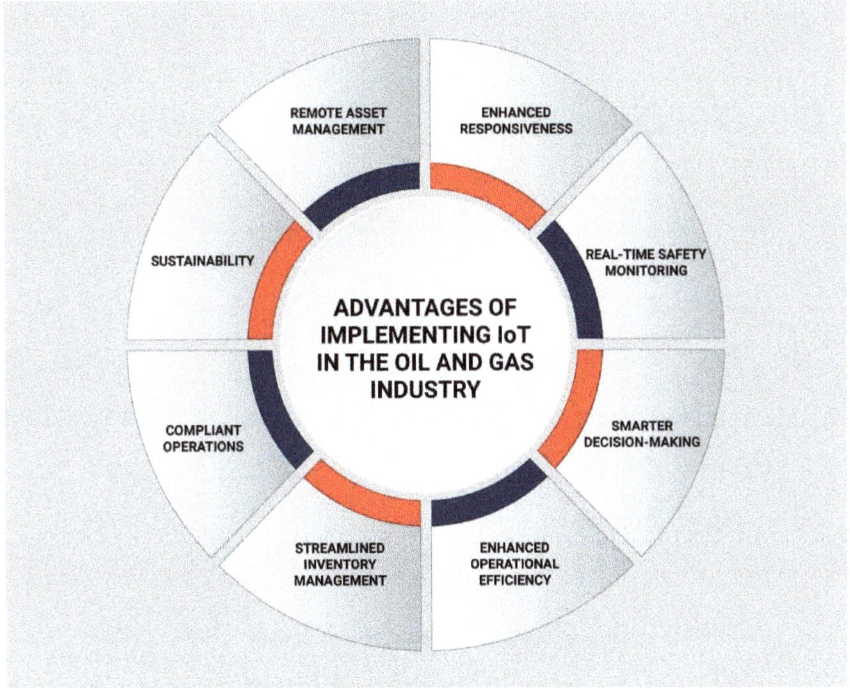

Figure 4.10 Some of the advantages of IoT in O&G [12].

4.7 CHALLENGES

The O&G industry has its share of challenges, including dealing with aging equipment and legacy systems, hazardous environments, stricter regulations, and geopolitical issues that hinder long-term business predictions. IoT brings forth a set of challenges that need to be addressed to ensure successful implementation. Other challenges of IoT applications in the oil and gas industry include the following [9]:

- *Data Security and Privacy:* As IoT systems collect, transmit, and store vast amounts of sensitive data, they pose significant risks to data security and privacy. IoT devices generate vast amounts of sensitive data, making them potential targets for cyberattacks. Ensuring the security and privacy of this data is a significant concern. Training employees on cybersecurity best practices is also essential.

• *Integration:* Many oil and gas companies have legacy systems in place, making it challenging to seamlessly integrate new IoT devices and platforms with existing infrastructure. Integrating IoT in oil and gas solutions with these existing infrastructures can be complex and time-consuming, requiring a clear understanding of the current systems and a well-thought-out implementation plan. One must ensure that IoT devices and platforms are compatible with industry-standard communication protocols.

• *Data Overload:* IoT devices generate massive volumes of data. Handling and processing this data can be overwhelming, leading to inefficiencies. Utilizing data analytics and machine learning algorithms can extract actionable insights from the data.

• *Scalability and Maintenance:* As IoT deployments grow, managing and maintaining a large number of devices and sensors can become complex and costly.

- *Talent Shortage:* IoT technologies require expertise in various fields, including data science, cybersecurity, and IoT device management. Finding and retaining skilled personnel can be a challenge. Investing in training and development programs for existing employees can build IoT expertise in-house. Collaborating with educational institutions and industry associations can create a pipeline of skilled professionals. Additionally, one may consider outsourcing specific IoT tasks to experienced service providers when necessary.

- *Regulatory Compliance:* The oil and gas industry is subject to stringent regulations related to safety, environmental impact, and data management. Adhering to these regulations while implementing IoT can be complex. An O&G company must engage with regulatory bodies and stay informed about evolving regulations. It should consider using IoT solutions that provide built-in compliance features to simplify reporting and monitoring.

- *Power and Connectivity:* Many oil and gas operations are located in remote or harsh environments, where power sources and connectivity may be limited. A company should explore low-power IoT devices and use alternative power sources such as solar or wind energy.

- *Investment Costs:* Deploying IoT solutions in the oil and gas industry often require significant upfront investments for hardware, software, and infrastructure upgrades. Companies must carefully evaluate these solutions' potential return on investment and long-term benefits before committing to implementation. Companies integrating IIoT solutions in the oil and gas industry are predicted to recover costs within three years of implementation of such advanced technologies.

4.8 CONCLUSION

The oil and gas industry is witnessing an unprecedented digital transformation, driven primarily by adopting Internet of things (IoT) technology. The integration of IoT technology in the oil and gas industry has emerged as a transformative force, revolutionizing operations, and enhancing efficiency across the entire value chain. Like many other industries, today the industrial Internet of things (IIoT) is the future of the oil and gas industry. Oil and gas have been confronting difficulties and is falling behind. As the cost of IoT hardware continues to decline, the oil and gas industry is becoming more enthusiastic about adopting digital technology. More information about IoT in O&G operations can be found in the books in [13,14] and the following related journals:

- *Petroleum*

- *Energy Reports*

- *Oil & Gas Journal*

REFERENCES

[1] "New technology trends for 2022," December 2021, Unknown Source.

[2] "How IoT app development is changing the oil and gas industry,"

https://blogs.emorphis.com/how-iot-app-development-is-changing-the-oil-and-gas-industry/

[3] "Breathtaking sunset at oil and gas complex,"

https://easy-peasy.ai/ai-image-generator/images/sunset-oil-gas-industry-complex-pipes-sky

[4] "7 Ways how IoT is revolutionizing the oil and gas industries,"

https://www.biz4intellia.com/blog/7-ways-how-iot-is-revolutionizing-the-oil-and-gas-industries/

[5] M. N. O. Sadiku, C. M. M. Kotteti, and J. O. Sadiku, "Internet of things in oil & gas industry," *International Journal of Trend in Scientific Research and Development,* vol. 11, no. 5, September-October 2024, pp. 126-131.

[6] A. Ghosh, "Internet of things: Basics," March 2014,

https://thecustomizewindows.com/2014/03/internet-things-basics/

[7] M. N. O. Sadiku, S. M. Musa, and S. R. Nelatury, "Internet of things: An introduction," *International Journal of Engineering Research and Advanced Technology*, vol. 2, no.3, March 2016, pp. 39-43.

[8] A. M. Rahmani et al., "E-Learning development based on Internet of things and blockchain technology during COVID-19 pandemic," *Mathematics,* vol. 9, 2021.

[9] "IoT for oil and gas industry: Advantages, features, and use cases," September 2023,

https://a-team.global/blog/iot-for-oil-and-gas-industry-advantages-features-and-use-cases/

[10] "Hype aside: Real-world use cases of artificial intelligence in the oil and gas industry,"

https://medium.com/instinctools/hype-aside-real-world-use-cases-of-artificial-intelligence-in-the-oil-and-gas-industry-a8c6f12fd10d

[11] A. Dziuba, "IoT in oil and gas: Transforming the sector with smart tech solutions,"

https://relevant.software/blog/iot-in-oil-and-gas/

[12] "How IoT in oil and gas industry is defining the future of energy," June 2023,

https://www.rishabhsoft.com/blog/iot-in-oil-and-gas-industry

[13] G. Blokdyk, *Internet of things in Oil and Gas.* 5STARCooks, 2022.

[14] R. F. Hussain et al., *IoT for Smart Operations in the Oil and Gas Industry: From Upstream to Downstream.* Elsevier Science, 2022.

CHAPTER 5

CLOUD COMPUTING

IN OIL AND GAS

"The cloud services companies of all sizes...The cloud is for everyone. The cloud is a democracy."

– Marc Benioff

5.1 INTRODUCTION

Digital transformation of oil and gas (O&G) companies is the integration of emerging technologies like artificial intelligence, machine learning, cloud services, automation, IoT, and data analytics across business functions. The goals of digitalization include improving efficiency, driving employee productivity, promoting sustainability, mitigating risks, and providing better customer and partner experiences. Figure 5.1 shows digitalization of the O&G industry [1].

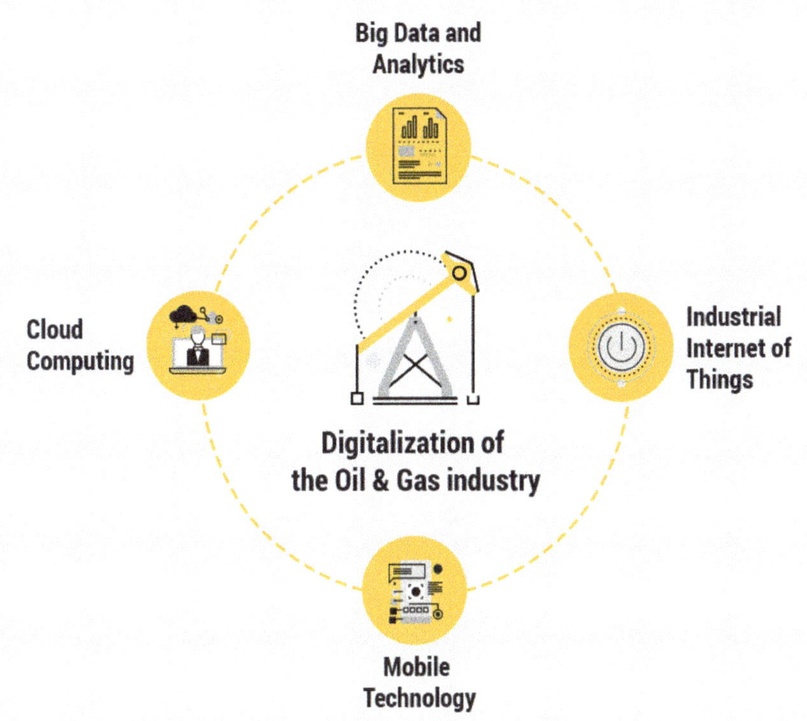

Figure 5.1 Digitalization of the O&G industry [1].

Cloud computing is a means of pooling and sharing hardware and software resources on a massive scale. Users and businesses can access applications from anywhere in the world at any time. Companies offering these computing services are called cloud providers and typically charge for cloud computing services based on usage [2]. Some features of cloud computing are displayed in Figure 5.2 [3]. The goal of cloud engineering is to create a highly scalable, flexible, and secure computing environment that can support the needs of modern businesses.

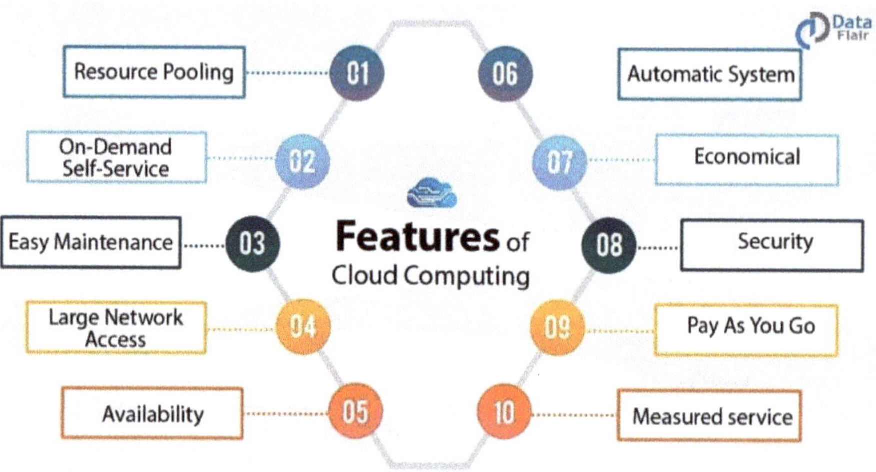

Figure 5.2 Some features of cloud computing [3].

Oil and gas (O&G) companies are businesses that explore, produce, refine, distribute, and sell oil and natural gas products. They have been around for over a century, but they have evolved dramatically in recent years. They operate in the energy sector and play a crucial role in meeting the world's energy demands. It is estimated that there are around 5,500 oil and gas companies globally. One of the major changes has been the integration of cloud computing technology into the O&G companies. These companies vary in size, from small local enterprises to large multinational corporations. Industry giants like Saudi Aramco, ExxonMobil, Shell, BP, and Chevron are at the forefront of adopting cloud technologies to streamline operations. The share of voice of the top 10 O&G companies in 2022 is depicted in Figure 5.3 [4].

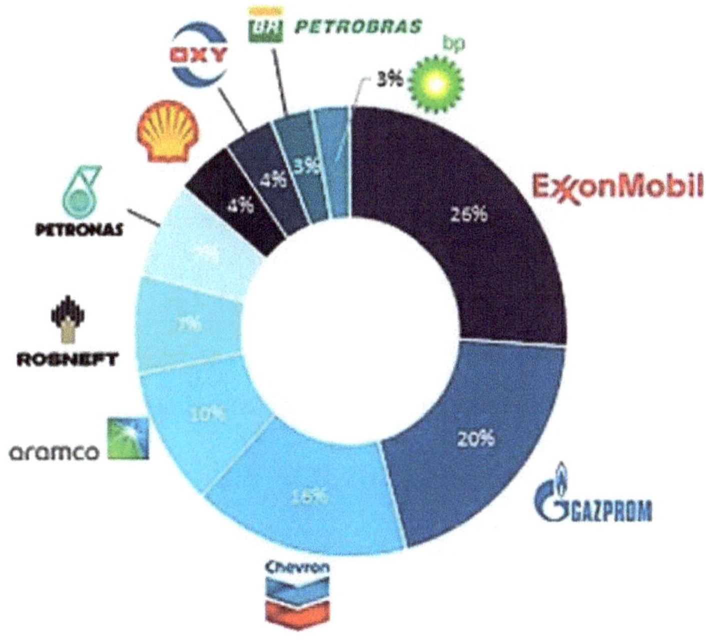

Figure 5.3 The share of voice of top 10 O&G companies in 2022 [4].

Cloud computing has become a key technology in the oil and gas industry, enabling companies to streamline processes, reduce costs, and drive innovation. It not only simplifies data analytics but also sets the stage for enhanced operational efficiency, thereby directly influencing the bottom line. With cloud computing, companies can essentially "rent" supercomputers as needed, without the upfront spend and maintenance associated with computers of that magnitude. For oil and gas professionals, cloud solutions allow more accurate modeling, data-backed drilling strategies, and optimized production [5].

This chapter delves into the integration of cloud computing within the oil and gas sector and how it has revolutionized the sector. It begins with explaining cloud computing basics. It describes cloud computing in oil and gas. It presents some applications of cloud computing in oil and gas. It highlights the benefits and challenges of cloud computing in oil and gas. The last section concludes with comments.

5.2 CLOUD COMPUTING BASICS

Cloud computing represents a newly emerging service-oriented computing technology. It is the provision of scalable computing resources as a service over the Internet. It allows manufacturers to use many forms of new production systems such as 3D printing, high-performance computing (HPC), industrial Internet of things (IIoT), and industrial robots. It is transforming virtually every facet of modern manufacturing. It is innovating, reducing costs, and bolstering the competitiveness of American manufacturing [6].

The key characteristic of cloud computing is the virtualization of computing resources and services. Cloud computing is implemented in one of three major formats: software as a service (SAAS), platform as a service (PAAS), or infrastructure as a service (IAAS). These services are illustrated in Figure 5.4 [7] and explained as follows:

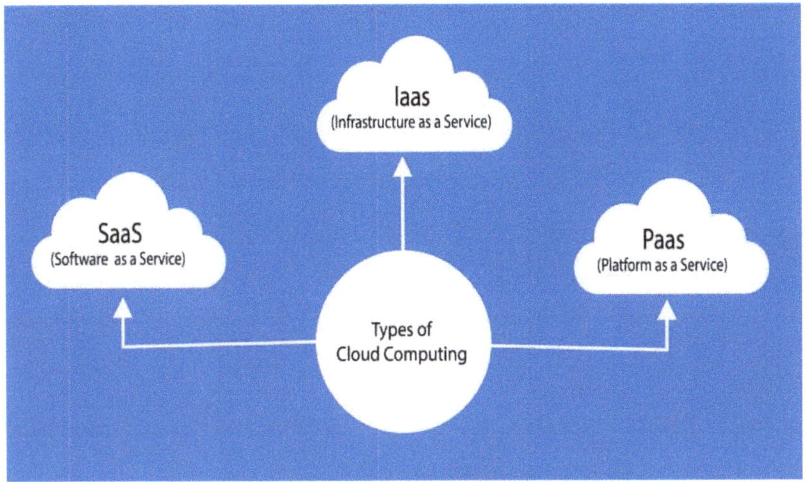

Figure 5.4 Three types of cloud computing [7].

SaaS: This is a software delivery model in which software and associated data are hosted on the cloud. In this model, cloud service providers offer on-demand access to computing resources such as virtual machines and cloud storage. Nowadays oil & gas companies transition to cloud computing and implement SaaS solutions for operations.

PaaS allows the end-user to create a software solution using tools or libraries from the platform service provider. In this model, cloud service providers deliver computing platforms such as programming and execution.

In the IaaS model, cloud service providers can rent manufacturing equipment such as 3D printers.

Just like cloud computing, CM services can be categorized into three major deployment models (public, private, and hybrid clouds) [8]:

- Private cloud refers to a centralized management effort in which manufacturing services are shared within one company or its subsidiaries. A private cloud is often used exclusively by one organization, possibly with multiple business units.

- Public cloud realizes the key concept of sharing services with the general public. Public clouds are commonly implemented through data centers operated by providers such as Amazon, Google, IBM, and Microsoft.

- Hybrid cloud that spans multiple configurations and is composed of two or more clouds (private, community or public), offering the benefits of multiple deployment modes.

These models are shown in Figure 5.5 [9]. Cloud computing finds application in almost every field.

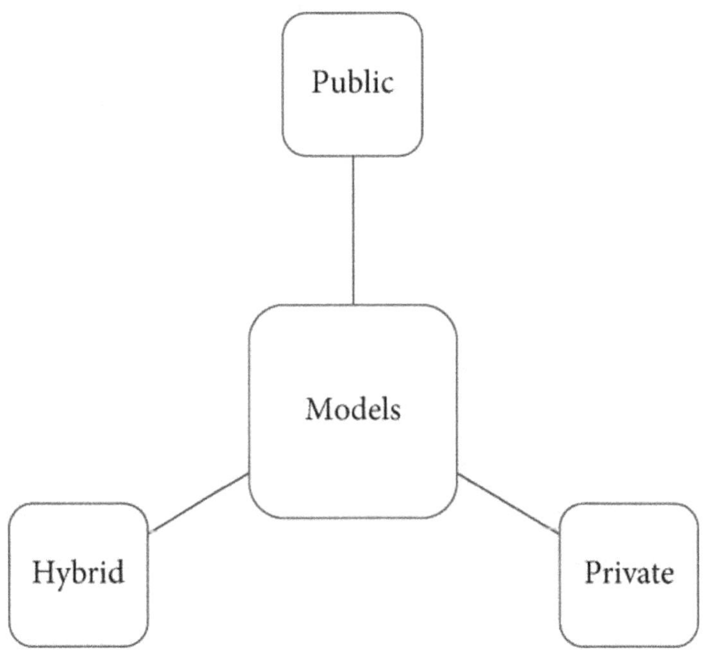

Figure 5.5 Cloud computing models [9].

5.3 CLOUD COMPUTING IN OIL AND GAS

The oil and gas industry encompasses a wide spectrum of activities ranging from exploration, extraction, and refining to the distribution of petroleum products. The industry is known for its complexity, as is typically shown in Figure 5.6 [10]. From upstream to downstream, oil and gas companies are turning to the cloud to reduce operating costs and streamline processes. If there is one commonality throughout the value chain of the O&G industry, it is the existence of unstructured operations, disparate eco-systems, disconnected processes, and disjointed efforts, and massive wastage of resources.

The oil industry is habitually alienated into three foremost components (upstream, midstream, and downstream), illustrated in Figure 5.7 [4] and explained as follows:

Figure 5.6 The oil and gas industry is known for its complex infrastructure [10].

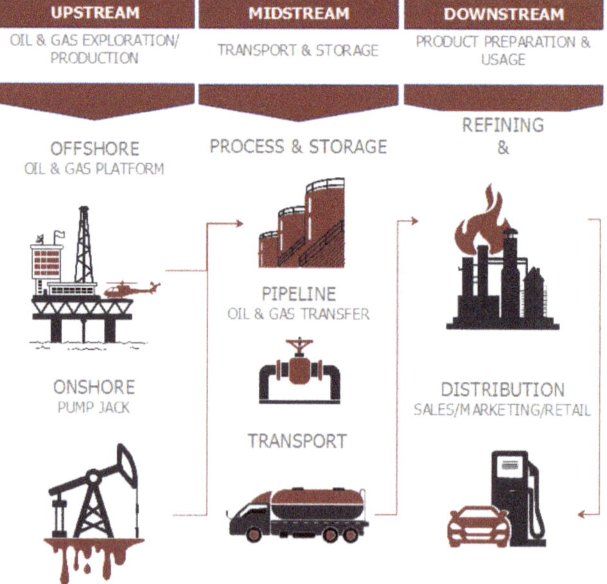

Figure 5.7 The oil industry has three components [4].

- *Upstream Sector:* The capital-intensive operations found in the upstream sector do not lend well to the trial-and-error approach typically used with new digital technologies. Upstream operators have collaborated with oilfield service companies for years to design products and processes to achieve drilling gains and obtain operational efficiencies. Machine learning and automation have applications across the upstream industry that can lead to percentage gains in production. In spite of compelling cases for cloud computing, the upstream petroleum industry faces technical challenges, notably reliance on massive datasets, ongoing legacy investments in IT, data security, and most of all, expertise.

- *Midstream Sector:* Midstream amenities follow gathering lines and transmission lines incorporating pumping station, control valve station, dehydration, fractionation, stockpiling, and other requirements based on the necessity of transportation on that location to carry oil to the consumers. The nature of pipeline transportation fills in as a national system to move the oil-based vitality assets from production regions or distribution ports to consumers, airplane terminals, military bases, and industry consistently. A pumping station is shown in Figure 5.8 [11].

Figure 5.8 A pumping station [11].

- *Downstream Sector:* Big data and cloud computing could affect the industry downstream, where hydrocarbon planning and marketing depend on supply, demand and consumer behavior. SaaS applications are useful throughout the oilwell lifecycle and also in downstream business applications.

5.4 APPLICATION OF CLOUD COMPUTING IN OIL AND GAS

The O&G industry plays a major role in the energy market and continues to influence the global economy as it produces the world's primary fuel. Cloud is one of the key technologies in this industry and remains the backbone for digital transformation. The technology enables O&G companies to move rapidly, show more agility and drive innovation. Cloud computing has impacted the oil and gas industry in several ways, including the following [12]:

- *Efficient Data Analysis* The oil and gas industry produces large volumes of data through its day-to-day exploration and production activities. The industry is gathering, storing, and analyzing its data on the cloud. Efficient analysis and timely insights into data can improve production by over 10%. The oil and gas industry is using cloud computing supercomputers to analyze all of the data being collected through devices and sensors. Supercomputers are faster and more powerful than standard computers and can analyze the massive amount of data coming in from oil and gas operations at record speeds. Cloud computing in the oil and gas industry allows companies to process this information quickly without spending the money upfront to purchase and manage these devices. More data analysis means more efficiency and less waste because it dampens the need for guesswork and roots out what does not work. The importance of real-time processing of data is shown in Figure 5.9 [13].

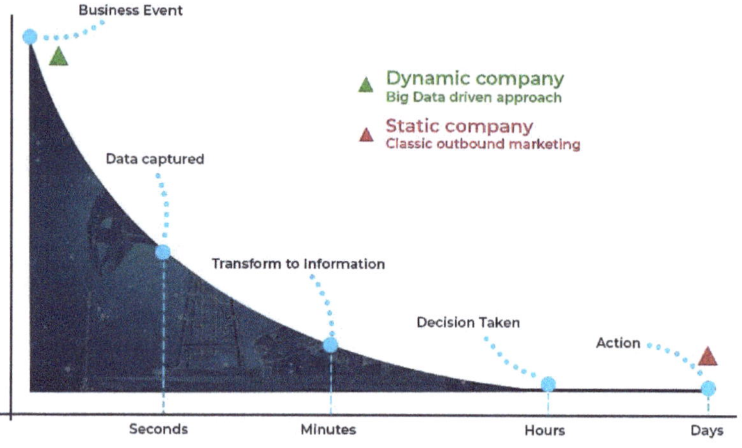

Figure 5.9 The importance of real-time processing [13].

- *Data Storage:* Storage hardware must evolve at a rapid pace to keep up with the incredible accumulation of data in today's world. Private servers used to be the answer for data storage, but private servers are extremely expensive to host and maintain. By contrast, cloud storage is scalable on the fly. Cloud computing allows oil and gas companies to scale their data management and storage, driving greater flexibility in infrastructure costs. The amount of data determines how much you spend on data storage. If the need for more storage arises, the cloud expands accordingly. A cloud company knows how to store, maintain, and expand its own storage locations better than a gas company. Outsourcing data storage to the cloud saves oil and gas companies the IT overhead cost they do not want to eat.

- *Automation:* Combining robotics with the cloud allows companies to establish the concept of "Robots-as-a-Service," or RaaS. Companies like BP, ExxonMobil, TotalEnergies, and even Ouro Negro deploy robotics regularly throughout their operations. From inspection to maintenance, detection, construction, cleaning, and heavy labor, robots are found at every level of energy production. Industrial robots are efficient; they can repeat a task thousands of times without changing or needing breaks, and they can operate in locations that may be unsafe for humans. Cloud programming and mechanized robotic technology allow companies to scale production to meet peak demands. Cloud robotics uses a rental business model but still allows firms to access the cloud to program, manipulate, and maintain robots—all remotely.

- *Remote Monitoring:* Cloud computing can enable remote monitoring and control of oil and gas assets, including wells, pipelines, and refineries. This can help improve oil and gas efficiency by reducing the need for manual intervention and enabling companies to monitor and optimize operations in real-time. Remote monitoring and automated operations can significantly reduce safety risks associated with traditional oil and gas operations. By leveraging remote monitoring solutions, you can enhance safety protocols, minimize human intervention, and mitigate operational risks.

- *Collaboration:* Digital transformation relies on the collaboration between people, processes, and technology. Oil and gas companies often work with geographically dispersed teams and must collaborate effectively. Companies must establish strong communication and collaboration between all team members in the cloud. Cloud-based collaboration tools can help improve communication and collaboration across the sector. Cloud computing allows for effective collaboration between stakeholders across departments through collaborative platforms for communication, reviews, and approvals. This simplifies workflow, documentation, process management, compliance, and decision-making, resulting in transparency, speed, and business growth. With today's complex energy assets, no single engineer can design an entire system alone. Collaboration with others is required.

- *Cost Reduction*: The pay-as-you-go model of cloud services eliminates the need for large upfront investments in on-premise hardware, leading to significant cost reductions. Cloud computing can reduce IT costs by up to 50%. In the oil and gas industry, cloud computing can help companies save costs by reducing the need for on-premises infrastructure, hardware, and software maintenance, as well as providing access to pay-as-you-go models that allow companies to pay only for the resources they use. Exploratory services rely heavily on data analysis and calculations. Storing and managing this data on-premise requires significant capital investment in hardware procurement, maintenance, upgrades, and IT personnel. Cloud technology significantly reduces these costs by allowing companies to pay only for the data storage space they use.

5.5 BENEFITS

The benefits derived from cloud computing are multifaceted. From boosting safety through robotics to facilitating precise drilling operations with real-time data analytics, the horizon of possibilities is expansive. Cloud computing has become a key technology in the oil and gas industry, enabling companies to reduce costs, improve efficiency, and drive innovation. Other benefits of cloud computing in O&G industry include the following [4]:

- *Security:* The journey towards cloud adoption is not devoid of skepticism, often rooted in concerns around data security and vendor lock-ins. Separate teams sharing responsibility for different parts of the cloud can create security risks if the information is not transferred between them correctly. In the oil and gas industry, data connectivity and security are crucial for safer and more successful drilling. Data connectivity and security make dangerous operations safer. The oil and gas industry is highly regulated and has numerous safety and security risks. Cloud computing can help improve safety and security by providing secure and compliant cloud infrastructure, disaster recovery solutions, and advanced security measures such as encryption and access controls.

- *Improved Efficiency*: The bottom line is that enhancing operational efficiency in the oil and gas industry by implementing automation technology and process optimization with the cloud will reduce risks but also help improve safety and lower overall costs. Cloud computing can help companies optimize operations from exploration to production. It can also help reduce downtime and improve asset utilization. Cloud computing in the oil and gas industry allows companies to create a more efficient and scalable structure. By monitoring and managing assets remotely through cloud-based systems, companies can identify potential problems early, minimize downtime, and improve overall operational efficiency.

- *Innovation:* Cloud computing can help companies accelerate their ability to innovate by enabling them to adopt the latest technologies, such as big data, IoT, AI, and machine learning.

• *Safety:* Cloud computing and robotics offer the oil and gas industry a chance to boost safety and efficiency by performing dangerous tasks remotely.

• *Real-time Insights:* Cloud computing is revolutionizing the oil and gas industry by turning vast data into actionable insights. It can help companies make smarter decisions in real-time.

• *Predictive Maintenance:* Cloud computing can help companies enable predictive maintenance and automate "smart" pipeline monitoring.

• *Predictive Analytics:* This enables oil and gas analysts to blend multiple data types and determine causes of failures. It can help operators recognize that a piece of equipment has a problem. Analysts are also able to scrutinize performance data to compare actual versus estimated costs.

• *Connected Operations*: Inefficiencies often arise as a result of complex processes and siloed operations. In the O&G industry, capital projects are executed by various stakeholders and functions, including suppliers, vendors, contractors, sub-contractors, and internal functions such as finance, procurement, and engineering.

- *Better Reliability:* Unplanned outages and downtime result in significant losses for O&G companies. Downtime can be caused by equipment failure, security breaches, or unidentified problems. Cloud computing can improve infrastructure reliability for O&G companies by offering predictive maintenance through data insights.

- *Lower Carbon Footprint:* Scientists and the UN have both stated that global emissions will need to drop by 7.8% per year in the next 10 years to prevent further temperature increases across the globe. Oil and gas companies are constantly under pressure to reduce their carbon dioxide and other greenhouse gas emissions. Virtualization can significantly increase server efficiency, resulting in less energy and electricity consumption. Additionally, cloud computing allows for servers to be utilized more efficiently and with less energy consumption.

- *Scalability:* The most significant benefit of the cloud computing eco-system is its ability to provide flexibility and scalability to adapt to the storage needs of the O&G company with agility and speed. With cloud computing, companies can scale their infrastructure up or down as needed, reducing costs and improving efficiency. Cloud infrastructure can be easily scaled up or down based on operational needs, providing flexibility to handle fluctuating data volumes during exploration, drilling, and production phases.

- *Monitoring*: Real-time monitoring also allows process design engineers and the operational teams to study usage patterns and predict potential failures.

5.6 CHALLENGES

The road to digital transformation is laden with challenges. The environmental impact of operations, marked by greenhouse gas emissions and pollution, along with the social implications, is a pressing concern. Oil and gas companies are facing several long-term challenges, including climate change and the scarcity of easily recoverable hydrocarbons. Amidst the challenges, cloud computing emerges as a beacon of solutions, which not only augment operational efficiency but also drive a substantial reduction in environmental footprint. Other challenges of cloud computing in O&G industry include the following [4]:

- *Air Pollution:* The extraction and processing of oil and gas release harmful pollutants like sulfur dioxide and nitrogen oxides into the air, which can have negative impacts on human health and the environment. Figure 5.10 shows typical air pollution due to oil and gas [4]. Cloud computing's potential to significantly reduce greenhouse gas emissions underscores its role as a cornerstone for a sustainable oil and gas industry. Cloud computing and advances in AI enable oil and gas companies to locate and prevent natural gas flare-ups and leaks which produce enormous amounts of carbon emissions.

Figure 5.10 A typical air pollution due to oil and gas [4].

• *Climate Change*: The burning of fossil fuels like oil and gas releases carbon dioxide into the atmosphere, which is a major contributor to climate change.

• *Water Pollution:* Oil spills and leaks from pipelines and offshore drilling rigs can contaminate water sources, which can harm aquatic life and negatively impact the ecosystem.

- *Habitat Destruction:* The exploration and development of oil and gas reserves can lead to the destruction of natural habitats and the displacement of wildlife.

- *Land Use:* Oil and gas drilling requires significant land use, which can lead to deforestation, soil erosion, and disruption of natural ecosystems.

- *Social Impacts:* The oil and gas industry can have negative social impacts on local communities, including displacement, land rights violations, and environmental injustice.

- *Increased Costs:* Multiple teams working on different parts of the cloud for oil and gas can increase costs. With so many people involved in various aspects of the cloud, it can be challenging to streamline processes and ensure everyone is working efficiently.

- *Integration Complexity:* O&G companies must confront the aging legacy systems still powering too many assets. Cloud-based platforms integrate IoT sensors, hardware, and software to enable remote, real-time asset monitoring. Integrating existing legacy systems with cloud-based platforms can be complex and require careful planning. Process industries like oil and gas are quite complex, with several types of equipment to track and manage in a facility.

- *Data management:* Collecting and managing more data can provide insights to improve efficiency and guide better decisions across operations. However, making the most of this data can be challenging.

- *Change Management:* The oil and gas industry constantly shifts as markets, politics, and technology evolve. Transitioning to a cloud-based environment may require significant changes in workflows and employee training.

- *Compliance:* As oil and gas operations go digital, companies need strategies to follow regulations and protect against hacking. With sites worldwide, companies must follow different rules, depending on their location and the work being done.

5.7 CONCLUSION

Cloud computing is a substantial lever propelling the O&G sector towards a future marked by sustainability, efficiency, and growth. The potential benefits of cloud computing in the O&G industry far outweigh the challenges, making it a crucial tool for driving innovation, cost reduction, and operational excellence in the industry. The industry can no longer afford to delay cloud adoption. In today's fast-paced digital landscape, the oil and gas industry is undergoing a profound transformation to stay competitive globally. This shift is affecting every aspect of the business, from exploration to distribution. It is a strategic move to thrive in the fourth industrial revolution. Today, cloud computing makes sense for small- to medium-sized oil and gas producers looking to develop a broader strategy toward effective digital innovation. It has become a vital component of the oil and gas industry, and its importance is only increasing. More information about cloud computing in the O&G sector can be found in the books in [14-16] and the following related journals:

- *Petroleum*

- *Petroleum Research*

- *Energy Reports*

- *Oil & Gas Journal*

REFERENCES

[1] "Digital trends for the oil & gas trends - Digitalization in oil and gas industry,"

https://www.kindpng.com/imgv/xoiwox_digital-trends-for-the-oil-gas-trends-digitalization/

[2] M. N. O. Sadiku, S. M. Musa, and O. D. Momoh, "Cloud computing: Opportunities and challenges," *IEEE Potentials*, January-February 2014, pp. 34-36.

[3] "Features of cloud computing – 10 major characteristics of cloud computing,"

https://data-flair.training/blogs/features-of-cloud-computing/

[4] M. Tyson, "Leveraging cloud computing for a sustainable oil and gas industry: A comprehensive insight," March 2023,

https://blog.brainboard.co/how-cloud-computing-is-revolutionizing-the-oil-gas-industry-fb12ffb03613

[5] M. N. O. Sadiku, P. O. Adebo, and J. O. Sadiku, "Cloud Computing in the oil and gas industry," *International Journal of Trend in Research and Development*, vol. 11, no. 6, November-December 2024, pp. 27-32.

[6] S. Ezell and B. Swanson, "How cloud computing enables modern manufacturing," June 2017

https://itif.org/publications/2017/06/22/how-cloud-computing-enables-modern-manufacturing

[7] "Cloud computing applications in agriculture,"

https://www.eescorporation.com/cloud-computing-applications-in-agriculture/

[8] "Cloud manufacturing," *Wikipedia,* the free encyclopedia

 https://en.wikipedia.org/wiki/Cloud_manufacturing

[9] J. S. Saini et al., "[Retracted] cloud computing: Legal issues and provision," *Security and Communication Networks*, vol. 2022,

https://www.hindawi.com/journals/scn/2022/2288961/

[10] "Future of oil & gas: Decline or endurance?"

https://netzero-events.com/the-future-of-oil-gas-predictions-and-pathways-for-the-next-decade/

[11] "Handy tips to leverage industrial IoT for oil and gas fleet management," July 2024,

https://www.intuz.com/blog/iiot-for-oil-and-gas-fleet-management

[12] "Cloud computing in the oil and gas industry: The benefits far outweigh the challenges," May 2023,

https://www.ctg.com/knowledge-center/blog/cloud-computing-in-the-oil-and-gas-industry-the-benefits-far-outweigh-the-challenges/

[13] "Cloud engineering in the oil and gas industry," April 2023,

https://www.linkedin.com/pulse/cloud-engineering-oil-gas-industry-revsolz#:~:text=Cloud%20engineering%20is%20a%20vital,the%20forefront%20of%20technological%20innovation.

[14] M. N. O. Sadiku, *Cloud Computing and Its Applications.* Moldova, Europe: Lambert Academic Publishing, 2024.

[15] M. M. Lawan, *Adoption of Cloud Computing Technology for Exploration, Drilling and Production Activities: Nigerian Upstream Oil and Gas Industry.* University of Wolverhampton, 2022.

[16] G. Cann and R. Cann, *Carbon, Capital, and the Cloud: A Playbook for Digital Oil and Gas.* Madcann Press, 2022.

CHAPTER 6

BLOCKCHAIN IN OIL AND GAS

"The whole point of using a blockchain is to let people — in particular, people who don't trust one another — share valuable data in a secure, tamperproof way."

– MIT Technology Review

6.1 INTRODUCTION

Important advances in technology rarely come with a full embrace at the start. Ground-breaking ideas can yield as many doubters as disciples. It takes time, experience, growth, and eventual acceptance to turn an idea into reality. Digital technologies are aimed at transforming the oil and gas supply chain services, reshaping industry competition and ensuring the safety of workers. Emerging technologies such as artificial intelligence, Internet of things (IoT), cloud computing, and blockchain can play a vital role in boosting the operational efficiency of the oil and gas industry. Recently, modern technologies have aided business titans in deciding to use digital currency rather than conventional trade in order to avoid fraud. Blockchain enables the existence of Bitcoin and many other cryptocurrencies. A cryptocurrency refers to a digital asset that works as a medium of exchange between various business organizations. Figure 6.1 shows the symbol of blockchain [1].

Figure 6.1 The symbol of blockchain [1].

The oil and gas industry encompasses thousands of companies spread across the globe. Both Chevron and ExxonMobil, two of the world's largest oil and gas companies, have shown interest in blockchain technology and its potential uses. In 2019, seven global firms in oil and gas companies formed the Global Oil and Gas Consortium. It is the first initiative of this kind. The recently announced consortium marks a groundbreaking milestone in the United States oil and gas industry. Chevron and ExxonMobil stand among the founding members [2].

Blockchain is a shared, distributed ledger which is to assist in recording the transactions and in trailing assets in a business network. The technology is already transforming industries and business processes, leading to completely new experiences and groundbreaking results, in terms of transparency, disruption, and innovation. It is increasingly becoming too compelling a technology that cannot be ignored for the oil & gas industry. It is emerging as a technology that demands attention within the oil and gas sector. It promises potential cost reductions and enhanced process efficiencies. It can aid contract execution in transactions where the level of counterparty trust is low or where transaction value or complexity is high. It assists in securing and simplifying oil and gas trading, shipment tracking, inventory control, documentation, and billing and payments [3].

In this chapter, we explore the potential opportunities and applications of blockchain technology in managing the exploration, production, and supply chain and logistics operations in the oil and gas industry. It begins with explaining what blockchain is all about. It describes blockchain in oil and gas industry. It presents some applications of blockchain in oil and gas. It highlights the benefits and challenges of big data in oil and gas. The last section concludes with comments.

6.2 WHAT IS BLOCKCHAIN?

Blockchain, a type of distributed digital ledger technology (DLT), is a relatively new and exciting way of recording transactions in the digital age. It is a decentralized and distributed digital ledger technology that securely records and verifies transactions across multiple computers or nodes in a network. Basically, it is a chain of blocks in which each block contains a list of transactions. The blockchain technology was created as the foundational basis for Bitcoin – a digital currency in which secure peer-to-peer transactions occur over the Internet. It is expected that the spending on blockchain solutions worldwide would grow from 4.5 billion USD (2020) to an estimated value of 19 billion USD by 2024 [4].

Originally developed as the accounting method for the virtual currency Bitcoin, Blockchains are appearing in a variety of commercial applications today. Blockchain technology is a type of distributed digital ledger that uses encryption to make entries permanent and tamper-proof and can be programmed to record financial transactions. It is used for the secure transfer of money, assets, and information via a computer network such as the Internet without requiring a third-party intermediary. It is now being adopted across financial and non-financial sectors. As a catalyst for change, Blockchain technology is going to change the business world and financial matters in major ways.

The first Blockchain was conceived in 2008 by an anonymous person or group known as Satoshi Nakamoto, who published a white paper introducing the concept of a peer-to-peer electronic cash system he called Bitcoin [5,6]. Bitcoin and Ethereum are the first two mainstream Blockchains. Other modern Blockchains include Namecoin, Peercoin, Ether, and Litecoin. Figure 6.2 shows different components of Blockchain [7].

Figure 6.2 Different components of blockchain [7].

Blockchain combines existing technologies such as distributed digital ledgers, encryption, immutable records management, asset tokenization and decentralized governance to capture and record information that participants in a network need to interact and transact. As illustrated in Figure 6.3, a complete blockchain incorporates all the following five elements [8]:

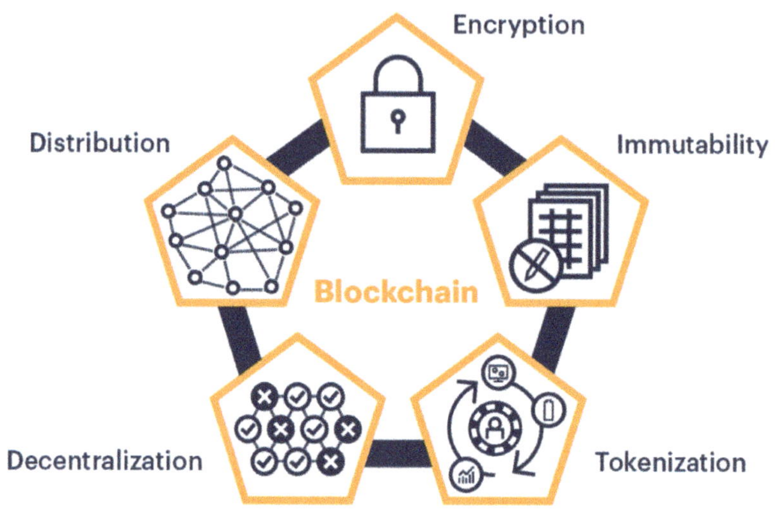

Figure 6.3 Five key elements of blockchain [8].

• *Distribution:* Digital assets are distributed, not copied or transferred. A protocol establishes a set of rules in the form of distributed mathematical computations that ensure the integrity of the data exchanged among a large number of computing devices without going through a trusted third party. A centralized architecture presents several issues including a single point of failure and problems of scalability.

• *Encryption:* BC uses technologies such as public and private keys to record data securely and semi-anonymously. Completed transactions are cryptographically signed, time-stamped, and sequentially added to the ledger.

- *Immutability:* The Blockchain was designed so these transactions are immutable, i.e. they cannot be deleted. No entity can modify the transaction records. Thus, Blockchains are secure and meddle-free by design. Data can be distributed, but not copied.

- *Tokenization:* Value is exchanged in the form of tokens, which can represent a wide variety of asset types, including monetary assets, units of data or user identities.

- *Decentralization*: No single entity controls a majority of the nodes or dictates the rules. A consensus mechanism verifies and approves transactions, eliminating the need for a central intermediary to govern the network.

Bitcoin and its underlying blockchain technology increasingly impact all facets of society. Bitcoin's status as digital gold is merely the tip of this technology. Figure 6.4 shows Bitcoin [9]. Although blockchain technology will for all time be associated with Bitcoin due to their common genesis, it has broader applications. Cryptocurrency will increasingly become a factor in family law issues as well.

Figure 6.4 Bitcoin [9].

For many businesses, blockchain is the road to transformation. The role of this technology in the oil and gas industry has evolved through various stages. The initial focus was on supply chain management and reducing paperwork. Subsequently, smart contracts were introduced to streamline transactions, and blockchain's cryptographic security features became crucial for data management and integrity. Leveraging blockchain's distributed ledger capabilities can reduce the amount of time spent reconciling price and volume differences among trade participants by making the same data available to all parties at the same time.

6.3 BLOCKCHAIN IN OIL AND GAS

The trading of oil and gas products such as gasoline and diesel is a highly standardized and quality-sensitive process that requires high security, privacy, and fast data processing. Oil & gas is a highly complex industry, sitting on the business of millions of barrels of oil and cubic meters of gas being bought and sold on international markets every day. The complex nature of the oil & gas industry involves multiple stakeholders and partners throughout the value chain. The complexity of the industry is depicted in Figure 6.5 [10]. The O&G industry does over $1.7 trillion dollars in business every year. A research study has estimated that the adoption of blockchain in the oil and gas industry can reduce the transaction execution time by 30%. Based on the type of service, oil and gas businesses are classified into upstream, downstream, and midstream categories [11,12]:

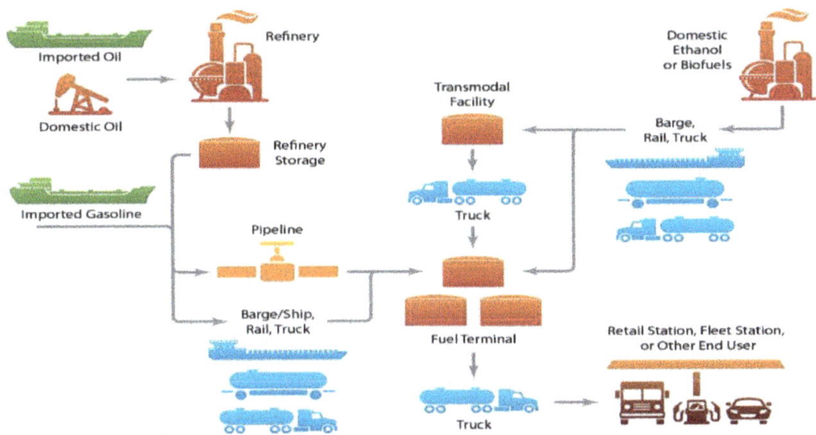

Figure 6.5 The complexity of the O&G industry [10].

- *Upstream Sector:* This is the section of the O&G industry that carries out resource exploration and extraction. The upstream service involves oil and gas exploration, extraction, or production of raw materials such as crude oil from wells. Other main activities include drilling contractors who own and operate drilling rigs, offshore oil/gas production facilities, oil service companies, and equipment manufacturers. The companies that carry out upstream oil and Gas processes operate under different contract requirements, each following its own performance metrics, and producing different deliverables. In upstream, there are too many equipment that are in daily use, which is very hard to keep track of and in consequence, there is a huge loss of time and money.

• *Midstream Sector:* Companies in this sector are responsible for storing and transporting the extracted crude oil and natural gas resources. The midstream sector involves (a) transportation services for safely shipping crude oil and natural gas using pipelines, tanker trucks, and heavy goods vehicles (HGV), (b) warehouse services for the temporary storage of crude oil at oil field terminals, and (c) refinement services to perform some initial processing of crude oil at the terminal. Storing and transporting hydrocarbon resources, managing expansive transportation networks, and paying dues to global regulations are what midstream oil and gas is all about. In midstream, there is a threat of faking transactions and contracts among third parties.

- *Downstream Sector:* This is the segment where the resources are refined and produced into final products which are sold to end-users at gas stations and other points of sale. The downstream stage involves the refinement of the crude oil (at the oil refining facility) shipped from various oil production sites to make useable and marketable products such as gasoline, petrol, diesel, and liquefied natural gas. When the product meets the consumer, companies will have a lot more tools to manage quality, provide diversified services, and maximize revenues with reduced expenses and accelerated payments. Blockchain solutions have the potential to shake up the industry and improve performance from the production stage all the way through to distribution and sales. In downstream, there is a cause for concern regarding data security and integrity.

Figure 6.6 shows the upstream to downstream workflow [13]. Many of the systems developed for automating business processes in the upstream, midstream, and downstream sectors have followed centralized architectures to store and process oil and gas-related data. The main challenges in each sector are, upstream: data leakage, midstream: data handling and replication, and downstream: integrity and data security. While the upstream companies typically face issues in heavy investment and cooperation among multiple parties, the other two sectors often have to deal with ownership changes, transactions, and information exchanges taking place across multiple companies. Together, they make an ideal candidate for blockchain application.

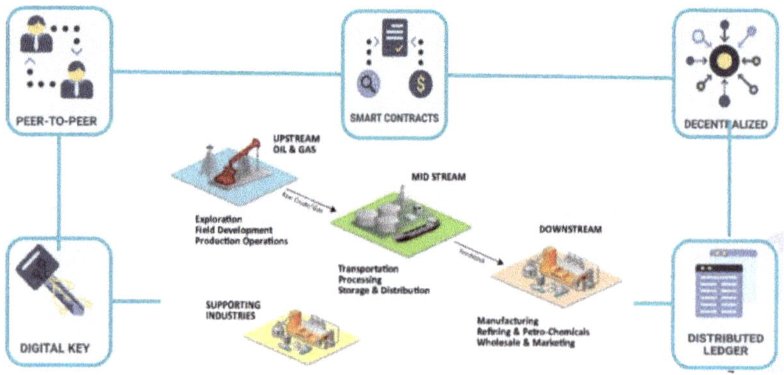

Figure 6.6 Upstream to downstream workflow [13].

6.4 APPLICATIONS OF BLOCKCHAIN IN OIL AND GAS

Blockchain is an emerging technology with many interesting use cases in oil and gas. It has a range of compelling applications within the oil and gas industry. It is an immutable digital ledger of economic transactions, that is secured through cryptographic methods and can be programmed to record the transaction of anything of value. Figure 6.7 shows blockchain use cases in the O&G sector [14]. Potential areas of opportunity include land transactions (by verifying and eliminating fraudulent land dealings), oil and gas sales (facilitating large transactions), complex sourcing (minimizing transaction inconsistencies), capital projects (adhering to contract terms), and joint ventures (improving cost and revenue-sharing audits). Various applications and opportunities of blockchain in the O&G industry are shown in Figure 6.8 [15]. Blockchain has a range of compelling applications within the oil and gas industry, including the following [16]:

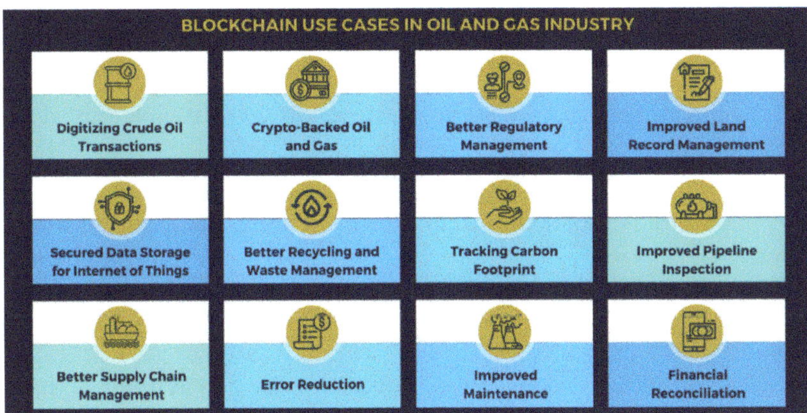

Figure 6.7 Blockchain use cases in the O&G sector [14].

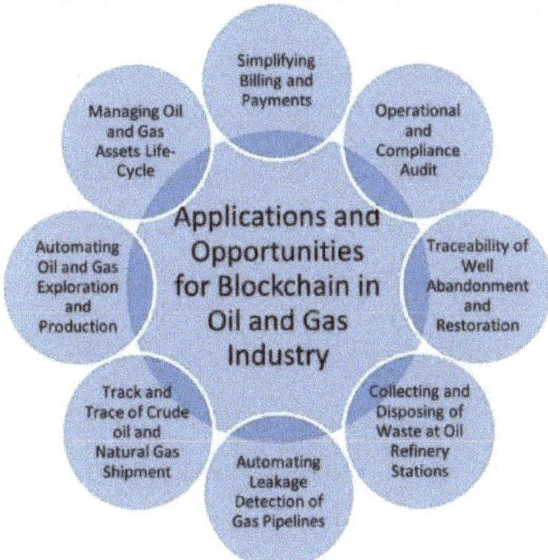

Figure 6.8 Applications and opportunities of blockchain in the O&G industry

[15].

• *Storage:* Blockchain can store accounting data and transactions directly on sensor devices, which can reduce process time. It can improve data storage for the Internet of things (IoT), which is used to monitor operations and increase efficiency. As the sector uses more sensor technology, blockchain can store transactions and accounting data directly on these devices, which can reduce process time by connecting assets directly to service contracts. As sensor technology reaches its peak within the industry, blockchain facilitates the direct storage of transactions and accounting data on these devices. The technology has the potential to accelerate digital transformation using sensors and cloud storage solutions. It can facilitate the direct storage of transaction and accounting data onto these sensors.

• *Blockchain Network:* Major oil & gas industry players are employing blockchain networks to drive forward unprecedented efficiencies with smart contracts, reduce market friction, and eliminate the need for intermediaries to carry out operations seamlessly. Each node in the blockchain network contains the same information, implying that once a transaction is committed on that network.

• *Contracts Execution:* Blockchain is quickly outgrowing from the crypto tag to become a smart contract facilitator. It facilitates embedding the contract in the transaction database for asset transfer. The contract is executed only once it is validated and deployed by all the parties. Smart contracts could add security and expediency to complex arrangements in the petroleum industry. They can introduce an additional layer of value by including timing and quality as factors in O& G deals. Blockchain can help with contracts that have high transaction values, are complex, or have low counterparty trust. By implementing smart contracts with distributed ledgers, operators can replace today's inefficient processes for handling payments, billings, statements and reports with one that is fast, reliable and transparent. To create a smart contract, terms are agreed upon by the entities involved, and the contract is then distributed on the blockchain. When the terms of the contract are fulfilled, the agreed-upon payment takes place automatically. Every contract would be distributed semi-anonymously across the entire network, making them safe from tampering or alteration. By using smart contracts and distributed ledgers to automate payments, well operators can simultaneously save money and improve their relationships with stakeholders.

- *Collaboration:* The oil and gas industry is a global leviathan, with a broad and diverse landscape where collaboration and open communication are the two main pillars. Blockchain can also transform contracting by providing a secure form of collaboration. It has the capacity to revolutionize contracting practices by offering a secure platform for collaborative endeavors. Implementing blockchain technology requires careful planning, collaboration with experts, and adherence to best practices.

- *Tracking:* Oil and Gas companies are now seeing the potential of blockchain technology to manage and track their resources more efficiently. Blockchain can facilitate real-time tracking of assets and enable sustainable practices like carbon credit trading. Tracking crude oil's path along the supply chain, as well as monitoring the conditions during storage, is crucial for companies to attain the highest quality of petroleum products. Personnel can easily keep track of the state of equipment using web apps connected to a decentralized network.

- *Hydrocarbon Tracking:* Blockchain technology can be used to track regulated substances effectively at each stage of the supply chain process. It can be utilized to track regulated substances like hydrocarbons viably at each phase of the supply chain process. This can help to improve accountability in the business. The exploration and production of hydrocarbons involve drilling wells that are spread across long distances, in locations with different jurisdictions, some of which are in geopolitically unstable regions, or even offshore sites. This can help improve accountability in the industry.

• *Supply Chain Transparency:* Oil & gas operations are highly complex due to their vast supply chain spread across the globe, involving physical side suppliers, producers, and distributors. This enormous supply chain footprint can ultimately slow down the operations, such as transaction approval, as it involves multiple stakeholders operating in different time zones. Blockchain and smart contracts, along with IoT devices, can change the traditional supply chain for oil & gas with the benefits of transparency and immutability. Transparency is achieved by the distributed storage and the embedded smart contracts, enabling blockchain to govern the entire business process management.

• *Land Transactions:* One of the major blockchain use cases in the oil and gas industry is improved land record management. This is the most critical aspect for oil and gas companies. It involves systematically managing land sale records, which are worth millions of dollars in investments. To record and validate each transaction, the platform uses blockchain. Blockchain can verify land ownership and eliminate fraud. It can also provide an audit trail of land transfers, which can help reduce disputes and title mismatches. Blockchain could allow oil companies to easily complete transactions throughout the entire energy supply chain. Blockchain can help facilitate large transactions. Fraudulent transactions would be impossible to fabricate, just as it would be impossible to hide a transaction, as both would require tampering with every entity on the network. One of the most prominent blockchain oil and gas use cases is the digitization process of crude oil transactions.

6.5 BENEFITS

Oil & gas companies that leverage blockchain can improve trade accuracy, increase scheduling and back-office efficiency, accelerate access to trade data, and shorten the working capital cycle. By using blockchain, companies can track materials and assets in real-time, save costs, and enhance collaboration. Blockchain assists in securing and simplifying oil and gas trading, shipment tracking, inventory control, documentation, and billing and payments. It can make processes more efficient, secure data, and create trust among different parties. Blockchain technology can be used in the oil and gas industry to improve efficiency, transparency, and accountability. It can optimize supply chains, reduce paperwork, and foster trust among stakeholders. Other benefits of blockchain in oil and gas include the following [11]:

- *Automation:* Like any other, the oil and gas industry has intermediaries, manual operations. Automation of such processes within the company and between contractors is the main goal of all services and departments. Blockchain can support the oil and gas supply chain by automating tasks and offering greater transparency.

- *Security:* This refers to the concerns that arise due to unauthorized access or attacks. The blockchain technology cannot be easily hacked, especially with numerous computational algorithms. The advantages of blockchain in the oil and gas industry manifest through enhanced transparency, compliance, and data security. The cryptographic security of blockchain ensures secure transactions, particularly beneficial for cross-border trades. Using blockchain, crude oil transactions can be digitized which ensures enhanced security, improved transparency, and optimized efficiency.

- *Privacy:* Beyond removing the middleman or intermediaries, the blockchain is now working as a trusted source where it promotes utmost privacy.

- *Trust:* Blockchain significantly boosts trust in the oil and gas industry. It ensures secure and transparent recording of data, like transactions and contracts, reducing fraud risk and improving efficiency. Blockchain technology has significantly impacted physical commodity trading by introducing transparency through its decentralized and immutable ledger, allowing stakeholders to track the supply chain and verify the origin and quality of commodities, thereby reducing fraud and enhancing trust. Blockchain's innovation promotes reliability and accountability, creating a more trustworthy and resilient oil and gas ecosystem. On top of boosting trust between companies and contractors/employees, such a blockchain network could also help cut down on hiring costs while ensuring improved job safety and performance.

- *Solutions:* As companies and investors across the oil and gas industry continue to operate in difficult times with compressed margins and reduced resources, blockchain provides real, difference-making solutions for the industry's most significant challenges today. Key benefits of the solutions include reduced cash cycle times, improved efficiency via lower overhead costs and fewer cost intermediaries, increased transaction visibility to help reduce the threat of tampering, fraud and cyber-crime, and the creation of transparent transactions by using shared processes and recordkeeping.

- *Transparency:* Blockchain is believed to have the potential to greatly impact the oil and gas industry by cutting down on operational time and costs while also introducing more transparency to the industry.

- *Decentralization:* One of the inherent qualities of a blockchain network is that it is decentralized, meaning it is not owned by a single entity, instead a network of computers (nodes) makes up a blockchain network. Blockchain can make this industry distributed and decentralized. Companies can make use of a decentralized network of IoT devices that trigger alerts and immediately notify corresponding personnel.

- *Error Reduction:* A major problem with human resources is the number of errors they make during data collection. Humans are prone to making errors, and even a single error in this regard can result in massive losses. However, as there is no way for this industry to get rid of all the human resources, it needs more technological help as it can get. In many cases, the companies can automate the data collection of maintenance data. This will significantly reduce the chance of errors.

- *Cost Reduction:* This refers to the cost reduction in operations when using blockchain technology. A decentralized blockchain system does not require a third party to take care of payment processes, unlike centralized financial infrastructures. So the transaction speed is increased, and in turn, the cost is significantly reduced.

- *Lack of Tools*: This refers to the lack of tools within an organization to use the technology. The tools could be necessary hardware and software to run the blockchain technology along with the maintenance. To implement widely, it could be expensive to invest in the companies.

- *Immaturity of Blockchain:* This refers to the part where the blockchain technology has not been used for some time and still has flaws. Blockchain technology is immature, and this immaturity causes technical difficulties such as scalability, usability, and interoperability.

- *Complex Sourcing*: Blockchain can help minimize transaction inconsistencies. It simplifies the unwieldy and complex oil and gas supply chain processes by introducing transparency to the involved business processes;

6.6 CHALLENGES

Blockchain in the oil and gas industry presents both challenges and opportunities. The industry now faces its most challenging times, against a backdrop of reduced consumer demand, scarce resources and funding, and the smallest margin for error due to harsh regulatory policies. Other primary problems for blockchain for oil and gas include scalability, as blockchain networks need to handle large volumes of transactions, and interoperability, ensuring different blockchains can communicate with each other. Other challenges include the following:

• *Regulation:* The O&G industry is among the most heavily regulated in the world with protocols deriving from various regulatory authorities from environmental to taxation. Federal law regulates the operation of natural gas pipelines that provide long-distance transmission and local customer distribution. Regulatory authorities will be able to maximize visibility in the industry as all the transactional data is stored on a blockchain network which can be accessed in real-time. Companies are struggling to comply with global regulations and to ensure that safety requirements are being met by every participant in the supply chain. Regulations can be enforced at every step in the technological process. Blockchain can help ensure compliance with contract terms.

• *Centralization:* A large portion of the existing systems is centralized, manual, and highly disintegrated which make them vulnerable to manipulation and the single point of failure problem.

• *Complex System:* Complexity is the degree of participation in environment coordination, and it is measured by the number of parties that collaborate to make an impact. If you add to the complexity of compliance and regulation regimes, the large number of projects involving a large number of contractors and subcontractors, you will have the world's most complicated document management system. The complexity of this process increases number of delayed payments and litigation and causes a persistent tendency to accumulate errors and inconsistencies. Upstream O&G process is a complex process every step of the way: from management to negotiating agreements and payment processing, to approving performance results and sending products further down the supply chain. Petroleum contracts are often complex and can involve many different entities.

• *High Cost:* Equipment in O&G sector costs from 125 to 500 thousand dollars, which in turn leads to an increase in the cost of goods for the final customer. Blockchain can improve audits for cost and revenue sharing. Companies are on a constant lookout for less costly and more effective ways of locating, extracting, and refining hydrocarbons to produce oil and gas products. Building new blockchain platforms and integration with the existing ecosystem of networks such as Ethereum can ultimately lower expenses and costs across the board.

• *Data Management:* The processes rely on manual entry, supervision, and are highly susceptible to security breaches and human error.

• *Water Management:* In the oil and gas industry, water management is a very complex and challenging issue due to the high volumes of water required, the hazardous nature of the wastewater, and the multitude of transactions of both clean and wastewater that need to be monitored. For oil and gas development operators, getting information about available source water is critical and time-consuming. In the water and wastewater context, the system of water flow is a network. Both the water and wastewater systems are decentralized networks.

• *Waste Management:* The horror of waste management in this industry is alarming. Most of the big companies do not have proper waste management protocols in place. These wastes are dangerously harmful to the environment and the people.

• *Inefficient Supply Chain*: Another huge problem in the O&G sector is the lack of proper supply chain management. In reality, every single oil or gas plant is huge in size and needs a lot of human resources to coordinate between all of the instances or elements.

- *Lack of Transparency:* Another huge problem is the lack of transparency in this industry. This is a huge burden the industry has to deal with, and it is not capable of stopping all the corruption. There is no transparency at every stage of the company.

6.7 CONCLUSION

Blockchain technology is worth considering for any application that requires a secure and transparent database and involves multiple parties. Because blockchain systems can simplify so many processes, large companies are developing infrastructure and software to support them. It has the potential to bring big changes to the oil and gas industry. The oil and gas industry uses state-of-the-art engineering solutions for oil and gas exploration but substantially lags behind in using innovative digital technologies. Although the applications of blockchain in the oil and gas industry have been promising, the actual implementation rate has remained low. Literature on the adoption of blockchain in the oil and gas industry is rare and success stories in the industry are rare. In spite of this, blockchain will be highly rewarding for oil and gas industry because it presents plenty of solutions for the industry. It also boosts efficiency and cuts unnecessary costs.

Careful analysis monitoring of evolving trends will tell if blockchain's future in oil and gas is transformative or transient. If the challenges that are present now in the field can be overcome, technology will have a bright future and will be able to change many aspects of the industry for a more technological future. The future of blockchain in the oil and gas industry is quite bright. More information about blockchain in the O&G sector can be found in the books in [17-20] and the following related journals:

- *Petroleum*

- *Petroleum Research*

- *Energy Reports*

- *Oil & Gas Journal*

REFERENCES

[1] "Blockchain meets the oil & gas industry," February 2018,

https://executiveacademy.at/en/news/detail/blockchain-meets-the-oil-gas-industry/

[2] "Blockchain technology in oil and gas industry," November 2023,

https://wezom.com/blog/blockchain-technology-in-oil-and-gas-industry

[3] M. N. O. Sadiku, P. O. Adebo, and J. O. Sadiku, "Blockchain in oil and gas industry," *International Journal of Trend in Research and Development,* vol. 11, no. 6, November-December 2024, pp. 20-26.

[4] C. M. M. Kotteti and M. N. O. Sadiku, "Blockchain technology," *International Journal of Trend in Research and Development,* vol. 10, no. 3, May-June 2023, pp. 274-276.

[5] "Blockchain," *Wikipedia,* the free encyclopedia

https://en.wikipedia.org/wiki/Blockchain

[6] S. Nakamoto, "Bitcoin: A peer-to-peer electronic cash system,"

 https://bitcoin.org/bitcoin.pdf

[7] "The beginning of a new era in technology: Blockchain traceability,"

https://www.visiott.com/blog/blockchain-traceability/#:~:text=The%20Beginning%20of%20a%20New,money%20without%20a%20central%20bank.

[8] "The CIO's guide to blockchain,"

https://www.gartner.com/smarterwithgartner/the-cios-guide-to-blockchain#:~:text=True%20blockchain%20has%20five%20elements,%2C%20immutability%2C%20tokenization%20and%20decentralization.

[9] "Helping the military adopt blockchain technology,"

https://www.pioneeringminds.com/helping-military-adopt-blockchain-technology/

[10] "Blockchain in oil and gas," February 2023,

https://graymatters-inc.com/blockchain-in-oil-and-gas/

[11] "Blockchain in oil & gas,"
https://www2.deloitte.com/us/en/pages/consulting/articles/blockchain-digital-oil-and-gas.html

[12] "Blockchain becomes the savior for oil and gas," August 2021,
https://pixelplex.io/blog/blockchain-oil-gas/

[13] "Blockchain solution (POC) for oil & gas industry and ERP integration,"
https://www.iterontech.com/ethereum-developer-oil-and-gas/

[14] H. Anwar, "Blockchain in oil and gas industry," January 2021,
https://101blockchains.com/blockchain-in-oil-and-gas-industry/

[15] "Blockchain in oil and gas industry: Applications and challenges,"
https://www.linkedin.com/posts/offer-thinktank_blockchain-in-oil-and-gas-industry-applications-activity-7226828273094127618-ntZE

[16] "Unleashing innovation in oil & gas industry with blockchain: 11 top use cases & benefits," October 2021,
https://www.birlasoft.com/articles/blockchain-oil-gas-industry-use-cases-applications

[17] M. N. O. Sadiku, *Blockchain Technology and Its Applications.* Moldova, Europe: Lambert Academic Publishing, 2023.

[18] U. Kishnani, N. Noah, and S. Das, *Blockchain in Oil and Gas Supply Chain*

A Literature Review from User Security and Privacy Perspective. SSRN, 2023.

[19] G. Blokdyk, *Blockchain in Oil and Gas a Clear and Concise Reference.* Emereo Pty Limited, 2018.

[20] S. Saraji and S. Chen, *Sustainable Oil and Gas Using Blockchain.* Springer, 2023.

CHAPTER 7

NANOTECHNOLOGY

IN OIL AND GAS

"Nanotechnology is manufacturing with atoms."

– William Powell

7.1 INTRODUCTION

Scientists and engineers have entered the enormous field of study known as nanotechnology to explore and improve material qualities at the atomic level. Successful applications of nanotechnology in many facets of life have boosted the effectiveness of product development and use. Nanomaterials are the building blocks of nanotechnology, and the impacts of the nanoscale are what give them their unique properties.

Over the past two decades, global energy consumption has been gradually increasing. Over the next 30 years, global energy demand is projected to rise as high as almost 60%, a challenging trend that may be met only by revolutionary breakthroughs in energy science and technology. Breakthroughs in nanotechnology open up the possibility of moving beyond the current alternatives for energy supply by introducing technologies that are more efficient and environmentally sound. Nanotechnology could be used to enhance the possibilities of developing unconventional and stranded gas resources. Nanotechnologist is concerned with building new structures and substances by manipulating molecules and atoms on this scale. In oil and gas applications, nanotechnology could be used to increase opportunities to develop geothermal resources by enhancing thermal conductivity, improving downhole separation, and aiding in the development of noncorrosive materials [1]. Research in nanomaterials has brought many performance enhancements to the materials used in oil and gas well drilling, cementing, production, and enhanced oil recovery (EOR). Research in petroleum well drilling, design, stimulation, and reservoir production management has demonstrated that nanoparticles can make industrial materials tougher [2].

Nanotechnology is the study of nanoscale phenomena, the practice of nanoscale engineering, and the use of materials at the nanoscale. It has become a ground-breaking applied technology in the past few decades. Several aspects of society have been impacted by nanotechnology, leading to increased productivity and lower-cost production of high-quality goods. Nanotechnology plays a significant role in the oil and gas industry by enhancing various processes. It is used for better oil recovery, real-time monitoring, innovative materials, drilling fluids, and reservoir characterization. Almost every oil & gas company is heavily investing in nanotechnologies to enhance oil recovery, to improve equipment reliability, to reduce energy losses during production, and to provide real-time analytics on emulsion characteristics [3].

This chapter explores how nanotechnology is transforming the oil industry and enhancing its performance. It begins with explaining what nanotechnology is all about. It covers nanotechnology in the oil and gas industry. It presents some applications of nanotechnology in oil and gas. It highlights the benefits and challenges of nanotechnology in oil and gas. The last section concludes with comments.

7.2 WHAT IS NANOTECHNOLOGY?

Technologies impact every aspect of our modern society. There are many ways in which our society and technology are interlinked. Nanotechnology has the potential to provide huge benefits, just like any useful technology.

The term "nano" means something small, tiny, and atomic in nature. The application of the term in science led to a field called nanotechnology. Nanotechnology refers to the characterization, fabrication and manipulation of structures, devices or materials that have one or more dimensions that are smaller than 100 nanometers.

It may be regarded as an area of science and engineering where phenomena that take place at the nano-scale (10-9m) are utilized in the design, production, and application of materials and systems. It is an emerging area that integrates chemistry, biology, and materials science to create new properties that can be exploited to gain new market opportunities [4].

Nanotechnology deals with the characterization, fabrication, and manipulation of biological and nonbiological structures smaller than 100 nm. Dimensions between approximately 1 and 100 nanometers are known as the nanoscale. As indicated in Figure 7.1 [5], the nanoscale is so small that we cannot see it with a light microscope. It is the scale of atoms and molecules. Nanotechnology involves the creation and application of materials and devices at the level of molecules and atoms. It may be regarded as the science that is conducted, researched, investigated, and experimented at the nanoscale. Nanotechnology is a multi-disciplinary field that includes biology, chemistry, physics, material science, and engineering. It is the science of small things—at the atomic level or nanoscale level. The past three decades have witnessed an increased interest and funding in nanotechnology. This has led to rapid developments in all areas of science and engineering [6].

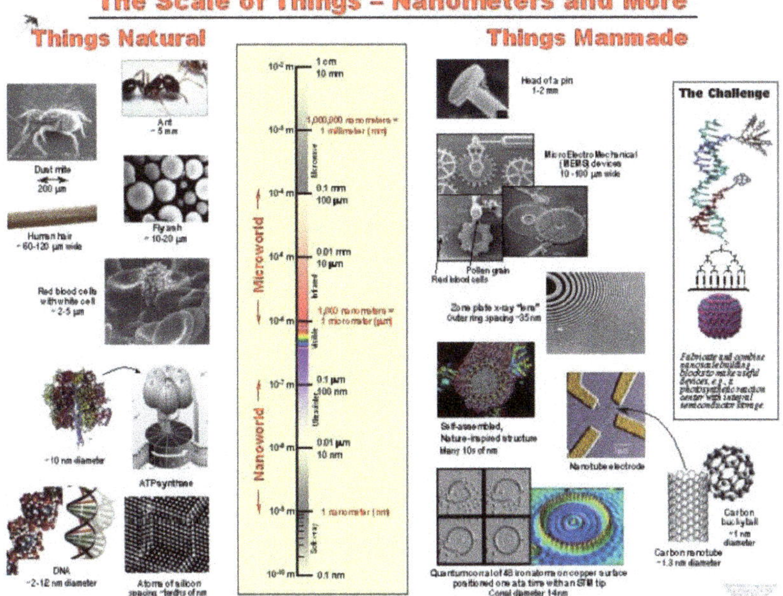

Figure 7.1 Indicating the relative scale of nanosized objects [5].

• Richard Feymann, the Nobel Prize-winning physicist, introduced the world to nanotechnology in 1959 and is regarded as the father of nanotechnology. Nanotechnology involves the manipulation of atoms and molecules at the nanoscale so that materials have new unique properties. Nanomaterials are expected to have at least one dimension (length, width, height) at the nanoscale of 1 – 100 nm. One nanometer is a billionth of a meter, too small to be seen with a conventional lab microscope. Nanomaterials include nanofilters, nanosensors, nano photocatalysts, and nanoparticles. Nanomaterials are known as nanoparticles when they have nanoscale length, width, and height. Figure

7.2 portrays a technique for the preparation of nanoparticles [7].

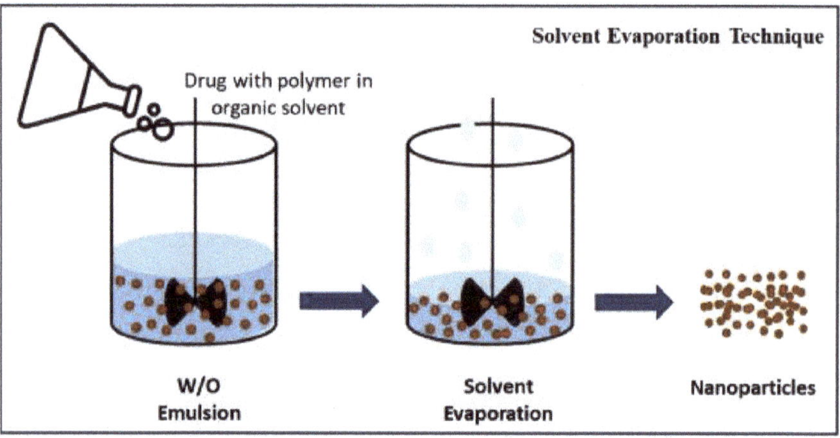

Figure 7.2 A technique for the preparation of nanoparticles [7].

Today, nanotechnology is part of our daily lives. Nanotechnology will leave virtually no aspect of life untouched. Its usages include everything from safer food processing to more efficient drug-delivery systems to powerful computer chips. Three steps to achieving nanotechnology-produced goods are [8]:

1.   Scientists must be able to manipulate individual atoms.

2.   Next step is to develop nanoscopic machines, called assemblers, that can be programmed to manipulate atoms and molecules at will.

3.	In order to create enough assemblers to build consumer goods, some nanomachines called replicators, will be programmed to build more assemblers.

Nanotechnology is trending among scientists and engineers. Here are some underlying trends one should look for [9]:

1.	Stronger Materials: The next generation of graphene and carbon devices will lead to even lighter but stronger structures.

2.	Scalability of Production: One big challenge is how to produce nanomaterials that make them affordable. Limited scalability often hinders application.

3.	More Commercialization: In addition to transforming the automotive, aerospace, and sporting goods fields, nanotechnology is facilitating so many diverse improvements: thinner, affordable, and more durable.

4.	Sustainability: One main goal of the National Nanotechnology Initiative, a US government program coordinating communication and collaboration for nanotechnology activities, is to find nanotechnology solutions to sustainability.

5.	Nanomedicine: There will be a mindboggling impact of nanotechnology on medicine, where advances are being made in both diagnostics and treatment areas.

Applications of nanotechnology are found in a wide range of industries, including engineering, medicine, microelectronics, manufacturing, biology, chemistry, energy, and agriculture, and life sciences. Figure 7.3 shows some applications of nanotechnology [10]. Although nanotechnology has been successfully applied in various industries, its use in the oil and gas sector is still limited.

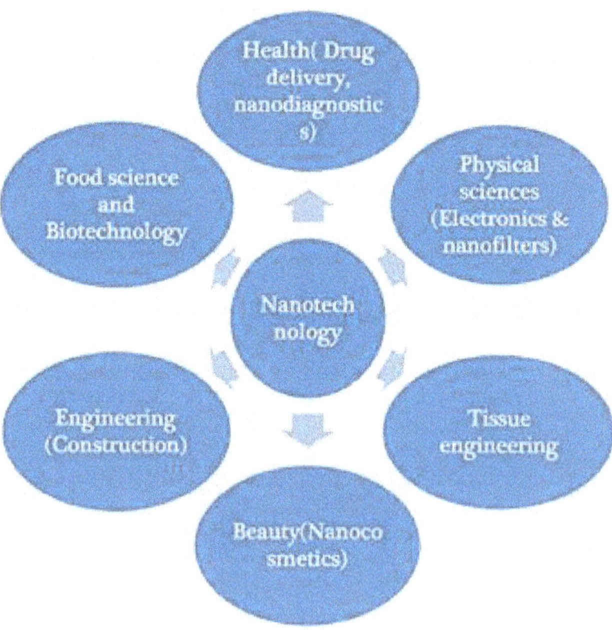

Figure 7.3 Some applications of nanotechnology [10].

In secondary and tertiary oil recovery operations, nanotechnology has the potential to significantly improve oil recovery and address problems caused by formation damage during water and gas reservoir flooding. Nanotechnologies can improve fluid phase separation, subsurface porous media qualities, coatings for reservoir components, and enhance the functionality of manufacturing system sensors and controls. Nanoparticles have gained significant attention in the oil and gas sector due to their unique properties and potential applications. Integrated circuit technologies that are lighter, with reduced size, and superior than ever before have been made possible by nanotechnology.

7.3 NANOTECHNOLOGY IN OIL AND GAS

Among the numerous energy sources, crude oil remains one of the most treasured, valued, and predominant natural resources on the planet Earth. Although demand for crude oil is increasing worldwide, discovering new oil reservoirs is difficult. Thus, researchers and oil companies must extract the remained oil in the matured reservoirs using efficient technologies. Nanotechnology is the current most attractive valuable technology [7].

Due to their special qualities, nanoparticles are useful in various applications in the oilfield, such as sensing or imaging, improved oil recovery (EOR), gas mobility control, drilling and completion, produced fluid treatment, and tight reservoir applications. Chemical processes have been facilitated in the oil and gas industries, from upstream to downstream, using materials with special size-dependent characteristics. Oil and gas field operations may be classified into two distinct sectors: upstream and downstream. The upstream sector pertains to the exploration and production of oil, while the downstream sector is primarily concerned with the refining, processing, purification, marketing, and distribution of the resulting products. The complexity of the O&G operation is shown in Figure 7.4 [11].

Figure 7.4 The complexity of the O&G operation [11].

Apart from the wide range of uses, such as structural nanomaterials, nanofluids, and nanosensors, nanotechnology also has prospective applications in upstream oil activities. The upstream oil business appears to have a promising future for nanotechnology if it can adopt notable breakthroughs from various industries. The upstream oil & gas industry could receive a great boost under the impulse of innovations in the field of nanotechnology.

7.4 APPLICATIONS OF NANOTECHNOLOGY IN OIL AND GAS

Nanotechnology has a wide range of uses, and its use in drilling fluids, innovative materials, real-time monitoring, reservoir characterization, well stimulation, cementing, wettability, enhanced oil recovery. and better oil recovery has shown tremendous promise. Figure 7.5 shows some of the applications of nanomaterials in O&G industry [12]. It is a rapidly growing field that has many applications in the oil and gas industry, including the following [13-15]:

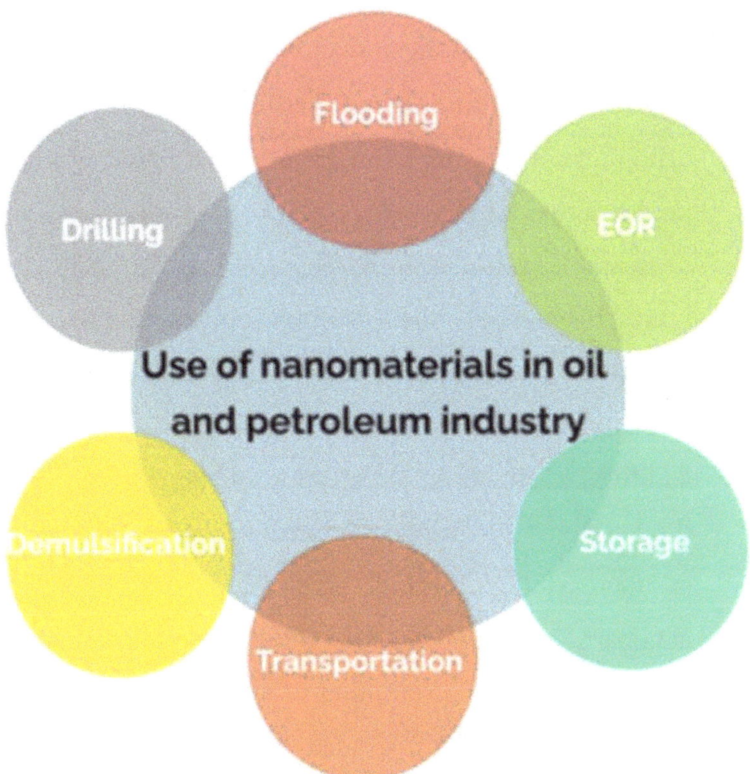

Figure 7.5 Use of nanomaterials in O&G industry [12].

- *Drilling:* The drilling process, as the name suggests, is creating a hole in the ground to reach the desired depth. Figure 7.6 shows a typical oil drilling [16]. Nanotechnology has shown promise in addressing major problems and improving drilling effectiveness. It has played a crucial role in the advancement of drilling tools and materials. Nanoparticles can improve the rheological properties of drilling fluids, which can help reduce fluid losses, stabilize the wellbore, and ensure safe drilling. They can also be used to create nano-drilling mud, which is often used in deep wells. Drilling fluids play a critical role in guaranteeing drilling success by improving oil recovery and cutting down on the time needed to achieve early oil production. They are comparable to blood in the physical makeup of the human body throughout the drilling operation. The primary function of additives is to improve the fundamental characteristics of drilling fluids. The variety of fluid additives available is a reflection of the complex drilling-fluid systems that are currently being used to enable drilling in increasingly difficult subsurface conditions. When nanoparticles are added to a fluid, the mixture becomes a nano-fluid. The incorporation of nanoparticles into drilling fluids has produced upgraded fluids with better features, including improved lubricity, thermal stability, and filtering capacities. Electrical and thermal conductivity are essential factors in drilling operations. It is necessary to use a thermally conductive drilling fluid to effectively remove heat from the drill bit.

Figure 7.6 A typical oil drilling [16].

• *Enhanced Oil Recovery* (EOR): This is a tertiary technique that involves the application of different methods for the purpose of increasing the amount of crude oil that can be extracted from a hydrocarbon reservoir. The use of nanoparticles in EOR is one of the most important fields of application as it provides larger amounts of oil during the extraction, thus ensuring a faster return on investment. Important EOR mechanisms include injection fluid viscosity increase, asphaltene precipitation and prevention, interfacial tension, and wettability changes. Different techniques using nanotechnology are being considered and very promising appears to be the use of nano-robots for real-time insight into the well pad. By adding some sensors inside the robots, very important information will be obtained. Nanoparticles can be used to change the surface tension and viscosity of reservoir fluids. They can also be used to create more efficient and eco-friendly chemical-enhanced oil recovery (CEOR) compounds. EOR could also be guaranteed by the use of nanoparticles dispersed in suitable fluids. One of the main problems in the oil & gas industry is the use of materials capable of withstanding highly corrosive environments. Different nanofluids are used in EOR techniques such as chemical, thermal, miscible, polymer, and microbiological flooding. Nanoparticles are added to fluids to create nanofluids. EOR techniques have been employed to extract the residual percentage of original oil from areas that are not amenable to water flooding. Figure 7.7 shows the categories of enhanced oil recovery technologies [7].

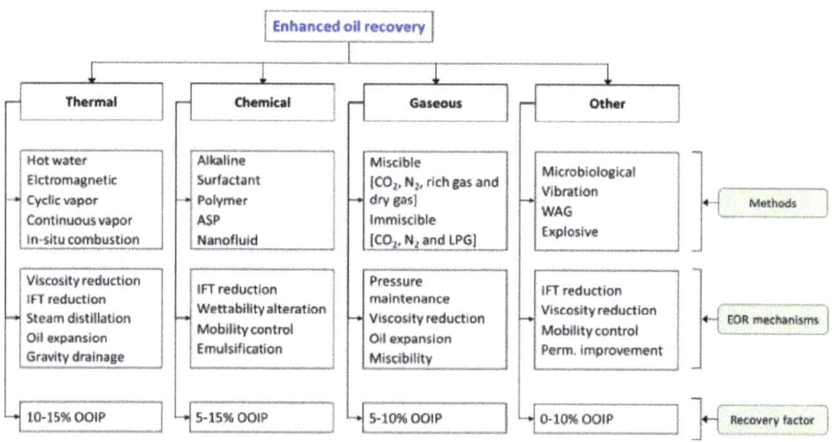

Figure 7.7 Categories of enhanced oil recovery technologies [7].

- *Oil Well Cementing:* Cement is needed in the oilfield for casing cementation in order to provide mechanical support, protection from corrosive fluids, and sealing. Cement set accelerators are essential in the cementing of oil wells by speeding up the transition from the liquid phase to the solid phase. Conventional cement-set accelerators, such as inorganic salts like calcium chloride ($CaCl_2$), are known to increase cement permeability while shortening the set time. The expenditures associated with the waiting on cement phase can be reduced through careful cement design. Nano silica has the potential to be used in a variety of oil and gas well cementing scenarios, such as cementing in lower temperature zones. Nano silica also addresses cement's problems with gas migration. The cement becomes more susceptible to pore pressures from the formation. Using a cement viscosity enhancer, using cement-set accelerators, regulating cement density, embedding nanosensors in cement, incorporating nanoparticle additives that promote microcrack self-healing, and controlling cement density will improve cement integrity.

- *Reservoir Characterization:* The characterization of reservoirs and improved oil recovery have benefited from nanotechnology. Improved technologies have been made available to characterize reservoirs through nanoparticle tracers and nanoscale imaging techniques. This has enabled the identification of the best drilling sites and increased productivity. While introducing nanoparticles into reservoirs, worries regarding the environmental impact surface.

- *Water Treatment:* Both upstream and downstream sectors employ a lot of water in accomplishing their activities, such as in drilling fluids, enhanced oil recovery/improved oil recovery (EOR/IOR), and in oil refinery processes. Treatment of contaminated water is essential but increasingly difficult due to the negative environmental effects of industrial water pollution. Using nanomaterials for water treatment and purification processes demonstrates appealing qualities. The nanofiltration separation process can be used to purify and desalinate injected water for EOR processes, water produced in oil fields, and water used in refineries. Comparing nanofiltration membranes are more efficient than the traditional technologies at eliminating pollutants from small adjourned oil droplets and dissolved substances. Water from the oil sands process, which is mostly produced during the oil sands production process, can also be cleaned using reverse osmosis and nanofiltration membranes.

- *Environmental Remediation:* One of the most significant predicaments confronting contemporary society is the pervasive environmental pollution and degradation stemming from various origins. Numerous conventional methods are being employed to tackle this predicament. One crucial area in which nanotechnology can have a significant impact is environmental stewardship, which addresses issues such as pollution, resource efficiency, and renewable energy production. Nanotechnology is presently being investigated for its potential deployment as a potent tool in combating environmental pollution. Nanosensors are essential for accurately tracking how human activity affects the environment and for enabling early intervention in the form of care, treatment, and prevention. Businesses that use nanotechnology have the ability to produce biodegradable, environmentally friendly materials, which could help reduce pollution. The utilization of nano remediation techniques presents a promising avenue for the detection and treatment of contaminants in various environmental matrices such as water, soil, sediment, and air. The nanoremediation technique encompasses the utilization of reactive materials to facilitate the detoxification and conversion of contaminants. It must be emphasized that this process is not only essential for the preservation of ecological well-being but also for the overall health of the general public.

- *Disinfection:* The drilling process for oil and gas can significantly enhance microbial activity due to the use of chemicals that are biodegradable and serve as a food source for microorganisms. This process poses a risk of introducing potentially harmful microbes to the environment. Nanotechnology is used to introduce a method of disinfecting the environment in order to combat the current and anticipated hazard. Nanomaterials possessing antimicrobial properties have the potential to be utilized for the purposes of disinfection and microbial control. The functionalization of carbon nanotubes has been shown to enhance their antimicrobial activity, thereby increasing their capacity for disinfection.

• *Desalination:* This process constitutes a mere 1 % of the global water consumption; yet it is an energy-intensive process, with the majority of operational expenses attributed to energy consumption. Given the significant impact of climate change, there is a pressing need to develop sustainable desalination processes that address the issues of brine discharge, greenhouse gas emissions, and energy consumption per unit of freshwater produced. Nanotechnology can play a crucial role in achieving specific energy consumption reduction.

• *Wettability*: This refers to a fluid's capacity to occupy a porous solid surface while other immiscible fluids are present. It is directly tied to how fluids interact with the solid or one another. The term "wettability alteration" describes how a surface's affinity for a specific phase changes. The upstream oil and gas sector finds wettability alteration (WA) of reservoir rock to be an appealing issue for increasing hydrocarbon production. Wettability modification is an essential component of improved oil recovery (EOR) procedures in the context of oil recovery. In order to improve the effectiveness of oil recovery, the wettability of the reservoir rock is altered from its original state. It has been observed that nanoparticles modify the wettability of the surface under study by altering the reservoir rocks' surface roughness. In other words, adding nanoparticles may increase surface roughness, which will reduce the contact angle and increase the system's wettability.

- *Oil Well Exploration:* Petroleum geologists are capable of conducting exploration, a process that involves utilizing various techniques to locate hydrogen resources beneath the earth's surface. In spite of the utilization of current sophisticated methods for oil recovery, such as thermal approaches, gas injection, water flooding, and chemical flooding, a significant amount of oil and gas may remain unrecovered. To overcome these technological challenges and effectively explore unconventional oil and gas sources, it is imperative to employ unconventional methods and exceptional materials. Cements in well construction are also witnessing numerous advances in the implementation of nanoparticle designs. The integration of nanoparticles and smart fluids can yield an exceptionally efficient sensor capable of operating effectively in demanding environments while delivering accurate measurements of temperature, pressure, oil flow rate, and stress in deep wells. Scientists are currently engaged in the development of nanosensors capable of performing reservoir characterization, identifying various fluid types, and monitoring fluid flow. Cement set accelerators are important chemical additives in oil well cements. They offer the possibility of shortening the time duration it takes for the cement to transition from liquid into the solid phase.

7.5 BENEFITS

Nanotechnology applications are rapidly expanding in various fields because of their unique qualities, such as a large surface area. It provides huge potential in addressing technology challenges related to the upstream oil and gas industry with nanomaterials and devices. Nanotechnology is becoming increasingly popular in the oil and gas industry. Other benefits include the following [13]:

- *Improved Performance:* Adding nanoparticles to drilling fluids can considerably improve their performance and solve some of their current problems. Different drilling fluid systems have shown the benefits of using nanoparticles in drilling fluids. Usually, drilling fluids are weighted to maintain a positive overbalance against the pore pressure of the formation. When designing drilling fluids for harsh conditions, particularly those with temperatures above 120 °C, nanoparticles have demonstrated promise.

• *Waste Management:* Solid waste materials encompass substances generated through human activities to fulfil their needs and subsequently introduced into the environment. Industrial and urban waste contribute hazardous organic and inorganic pollutants to water, soil, and air. Conventional technologies struggle to effectively eliminate these pollutants. Common biological and physicochemical methods are unable to remove all of the toxic and nonbiodegradable materials found in waste. Modern technologies, such as nanotechnology, emerge as a crucial solution to address this issue.

• *Increased Productivity:* Through increased productivity and superior products created at lower overhead costs, which boosts demand, nanoparticles have had a clear and significant impact on society in many fields. Nanoparticles are therefore crucial to contemporary lifestyles. Nanotechnology has shown that nanoparticles outperform similar macro- and micro-materials in terms of their chemical, physical, thermal, mechanical and tribological capabilities.

- *Sensing and Imaging:* Nanoparticles can be used to create temperature and magnetic sensors, as well as imaging techniques. For example, magnetic nanoparticles can be used as contrast agents to monitor and surveil reservoirs.

- *Miniaturization:* Some of the possible benefits of nano-materials are the outcome of miniaturization, while others are the result of the change in the property of the material.

7.6 CHALLENGES

The widespread adoption of nanoparticles in the oil and gas field is impeded by several challenges that require attention. These challenges include cost, durability, compatibility, safety, and regulation. A major challenge in the oil & gas industry is the use of materials capable of withstanding highly corrosive environments. Nanotechnology is still in the early stages of application in the oil and gas industry, and there are some challenges to overcome. Other challenges include the following [13,15]:

- *Temperature:* Most recently discovered oil and gas formations are situated in high-pressure and high-temperature conditions. Exploration and production under such environments are costly and challenging. The rheology of drilling fluids is considerably impacted by the elevated temperatures. For instance, clay swelling, flocculation, and sodium ion replacement in bentonite mud, frequently used in drilling, cause increased yield stress at higher temperatures. High temperatures can also destroy polymeric additives in drilling fluids, diminishing viscosity and performance and creating new difficulties for drilling operations. Nanomaterials must be able to withstand the high-temperature and high-pressure conditions of the downhole environment. Deep oil wells and steam injectors have difficulties due to uncontrolled heat loss, which can result in problems including clogging, the deposition of paraffin and asphaltenes, and poor steam quality. For example, the strength of cement can decrease dramatically at temperatures above 150 °C.

• *Environmental Impact:* In commercial drilling, managing drilling fluid waste is a major environmental concern, and actions are being taken to mitigate the effect of drilling activities utilizing non-aqueous fluids on the metal content of marine sediments. It is common knowledge that drilling mud pits at drilling rigs can contain sizable concentrations of hazardous heavy metals. Nanotechnology can make the O&G industry considerably greener.

- *Injection of Nanofluids:* The continual injection of nanofluids into the water stream is, in fact, a major problem in the enhanced oil recovery (EOR) process. This raises concerns over potential injectivity loss in the injector well due to numerous types of formation damage. Consequently, it is imperative that the assessment of flow assurance in the laboratory is given top priority to avoid and minimize any formation damage related to nanoparticle injection.

- *Collaboration:* Nanotechnology is characterized by collaboration among diverse disciplines. It is crucial to promote cooperation between regulatory agencies, businesses, and academia to guarantee the appropriate and long-term integration of nanotechnology in the oil field. A thorough examination of potential hazards and environmental effects, as well as the formulation of rules and standards for the use of nanomaterials, would all be made easier by collaboration.

- *High Cost*: The production and use of nanoparticles can be expensive, thereby limiting their commercial adoption in the oil and gas industry. The application of nanotechnology is limited due to the fact they are so expensive to produce and o   maintain. The price of nanoparticles continues to be an important factor, and lowering it will be essential for their widespread use and commercialization in the industry.

• *Degradation:* Nanoparticles are susceptible to degradation in the harsh downhole environment, which can limit their effectiveness and longevity. Using nanoparticles for EOR requires them to travel deep into the reservoir and remain stable in the harsh downhole environment.

• *Incompatibility:* Achieving compatibility between nanoparticles and other fluids and materials used in the oil and gas industry can be challenging, particularly for nanoparticles designed for specific applications.

• *Safety*: In spite of the numerous potential benefits of incorporating nanoparticles into the oil and gas sector, concerns regarding their safety and environmental impact persist. Nanomaterials and their biodegradability have raised concerns regarding long-term consequences and safety. Nanoparticles can pose a safety hazard if not handled properly, as they can be easily inhaled or ingested. It is necessary to proceed with caution and conduct additional research to fully comprehend both their potential benefits and risks. Some of the most widely studied nanomaterials in the oil and gas sector are nano silica and metallic oxide nanoparticles; they have various effects on human beings and the environment.

- *Regulation:* The use of nanoparticles in the oil and gas industry is a relatively new, and there are few regulations governing their use, leading to uncertainty and delays in commercialization.

7.7 CONCLUSION

Nanotechnology has stood strong in the oil and gas industry, with many applications that have gone from laboratory and simulation studies to successful trial applications in the field. It will continue to influence the future of the oil and gas sector by embracing ongoing research, development, and responsible application, enhancing drilling operations, boosting sustainability, and fostering overall industry advancement. Prominent oil corporations are currently engaged in a thorough examination of the potential implementation of nanoparticles and nanotechnology within the oil and gas sector. The oil and gas field is expected to see a successful expansion in the application of nanotechnology in different areas. More information about nanotechnology and nanomaterials in oil & gas industry can be found in the books [17-23] and the following related journals/magazines:

- *Nanotechnology*

- *Nanoscale.*

- *Nano: The Magazine for Small Science*

- *Micro and Nano Technologies*

- *Nanotechnology News*

- *Nature Nanotechnology*

- *Current Research in Nanotechnology*

- *American Journal of Nanotechnology & Nanomedicine*

- *Nanomedicine: Nanotechnology, Biology and Medicine*

- *Journal of Nanotechnology*

- *Journal of Nanoparticle Research*

- *Journal of Bioelectronics and Nanotechnology*

- *Journal of Nanoscience and Nanotechnology,*

- *Journal of Micro and Nano-Manufacturing*

- *Journal of Nanoengineering and Nanomanufacturing*

- *Nanotechnology and Precision Engineering*

- *South African Journal of Science*

- *International Journal of Petroleum and Petrochemical Engineering*

- *Journal of Petroleum Science and Engineering*

- *Petroleum*

- *Petroleum Research*

- *Energy Reports*

- *Oil & Gas Journal*

REFERENCES

[1] A. N. Jawahar, "Applications of nanomaterials and nanotechnology in oil and gas industry," *International Journal of Petroleum and Petrochemical Engineering* (IJPPE), vol.6, no. 1, 2020, pp. 28-32.

[2] P. J. Boul, and P. M. Ajayan, "Nanotechnology research and development in upstream oil and gas," October 2019,

https://onlinelibrary.wiley.com/doi/full/10.1002/ente.201901216

[3] M. N. O Sadiku, M. Oteniya, J. O. Sadiku, and S. Abunene, "Nanotechnology in oil and gas industry," *International Journal of Trend in Scientific Research and Development*, vol. 8, no. 5, September-October 2024, pp. 842-852.

[4] M. N.O. Sadiku, M. Tembely, and S.M. Musa, "Nanotechnology: An introduction," *International Journal of Software and Hardware Research in Engineering,* vol. 4, no. 5, May. 2016, pp. 40-44.

[5] "Nanotechnology white paper,"

https://www.epa.gov/sites/default/files/201501/documents/nanotechnology_whitepaper.pdf

[6] M. N. O. Sadiku, Y. P. Akhare, A. Ajayi-Majebi, and S. M. Musa, "Nanomaterials: A primer," *International Journal of Advances in Scientific Research and Engineering,* vol. 7, no. 3, March 2020, pp. 1-6.

[7] U. A. Ali et al., "A state-of-the-art review of the application of nanotechnology in the oil and gas industry with a focus on drilling engineering," *Journal of Petroleum Science and Engineering,* vol. 191, August 2020.

[8] K.R. Saravana and R. Vijayalakshmi, "Nanotechnology in dentistry," *Indian Journal of Dental Research*, November 2005.

[9] N. S. Giges, "Top 5 trends in nanotechnology," March 2013,

https://www.asme.org/topics-resources/content/top-5-trends-in-nanotechnology

[10] D. E. Effiong et al., "Nanotechnology in cosmetics: Basics, current trends and safety concerns—A review," *Advances in Nanoparticles*, vol. 9, 2020, pp. 1-22.

[11] Bottom of Form "The role of nanotechnology in the oil and gas industry," May 2018,

https://nano-magazine.com/news/2018/5/30/the-role-of-nanotechnology-in-the-oil-and-gas-industry

[12] A. Roy et al., "A review of nanomaterials and their applications in oil & petroleum industries," *Nano Express*, vol. 4, 2023.

[13] A. M. Alkalbani and G. T. Chala, "A comprehensive review of nanotechnology applications in oil and gas well drilling operations," *Energies,* vol. 17, no.4, 2024.

[14] A. Milioni, "Nanotechnology in oil industry," June 2015,

https://www.oil-gasportal.com/nanotechnology-in-oil-industry/

[15] M. Emmanuel, "Unveiling the revolutionary role of nanoparticles in the oil and gas field: Unleashing new avenues for enhanced efficiency and productivity," *Heliyon*, vol. 10, no. 13, July 2024,

https://www.sciencedirect.com/science/article/pii/S240584402409
9882

[16] "What is oil drilling: Everything you should know,"

https://www.linedpipesystems.com/oil-drilling/

[17] M. N. O. Sadiku, S. M. Musa, T. J. Ashaolu, and J. O. Sadiku, *Applications of Nanotechnology*. Gotham Books, 2023.

[18] T. A. Saleh, *Nanotechnology in Oil and Gas Industries: Principles and Applications*. Springer, 2018.

[19] A. Hassan, and S. Atta, R. Nabaz, *Nanotechnology In Oil And Gas Industry: Design And Manufacture Of Advanced Nano-Material*. LAP LAMBERT Academic Publishing, 2019.

[20] C. Huh et al., *Practical Nanotechnology for Petroleum Engineers*. Boca Raton, FL: CRC Press, 2019.

[21] N. S. El-Gendy, H. N. Nassar, and J. G. Speight, *Petroleum Nanobiotechnology: Modern Applications for a Sustainable Future*. Apple Academic Press, 2022.

[22] T. A. Saleh, *Applying Nanotechnology to the Desulfurization Process in Petroleum Engineering*. IGI Global, 2015.

[23] T. Sharma, K. R. Chaturvedi, and J. Trivedi, *Nanotechnology for CO2 Utilization in Oilfield Applications*. Elsevier Science, 2022.

CHAPTER 8

3D PRINTING IN OIL AND GAS

"3D printing represents the democratization of manufacturing. It allows anyone to create anything, anywhere, at any time."

– Neil Harbisson

8.1 INTRODUCTION

The oil & gas (O&G) industry is one of the largest industries in the world and it is a critical part of the energy sector. It is the number one energy source, especially when it comes to heating. To reduce wasteful consumption and accidents due to leaking, oil & gas companies are turning to 3D printing (aka additive manufacturing) to create geometrically complex, cost-effective parts [1]. These include companies such as Shell, AML3D, ExxonMobil, Chevron, BP, Siemens Energy, Halliburton, Total, and General Electric. These major companies have published stories about their use of 3D printing for prototyping and production applications in their industry. Also, the US Department of Energy (DOE) has been supporting the development of 3D printing for applications in the energy sector. In July 2018, the department selected some projects for R&D to develop innovative technologies for fossil fuel power systems.

The advent of 3D printing technology has emerged as a game-changer in numerous industries, completely transforming conventional methods of production. The oil and gas industry, known for its complex and demanding operations, is one of the major untapped markets for additive manufacturing although it is important for the industry as it can help in boosting productivity and reduce operational costs. According to the World Economic Forum, Additive manufacturing can make oil and gas companies save cost and time worth up to $30 billion.

Oil and gas industry is one of the biggest industries in the world in terms of dollar values. Companies in the oil and gas industry influence the global economy by engaging in the exploration, extraction, refining, and transportation of one of the primary fuel sources. As oil and gas production needs increase, so too do operators' demand for efficiency and sustainability. Additive manufacturing (or 3D printing) is a valuable tool that enables the faster and more efficient manufacturing of spare parts, equipment, and components while reducing downtime and maintenance costs. 3D-printed parts are being utilized by the energy sector to harness the natural resources of our planet [2].

This chapter aims to highlight significant opportunities and challenges relating to the adoption of 3D printing in the oil and gas industry. It begins by explaining what 3D printing is all about. It describes different types of 3D printing. It covers 3D printing in the oil and gas industry. It presents some applications of 3D printing in oil and gas. It highlights the benefits and challenges of 3D printing in oil and gas. The last section concludes with comments.

8.2 WHAT IS 3D PRINTING?

3D printing (also known as additive manufacturing (AM) or rapid prototyping (RP)) was invented in the early 1980s by Charles Hull, who is regarded as the father of 3D printing. Since then it has been used in manufacturing, automotive, electronics, aviation, aerospace, aeronautics, engineering, architecture, pharmaceutics, consumer products, education, entertainment, medicine, space missions, military, chemical industry, maritime industry, food industry and jewelry industry. All the parts created using a 3D printer need to be designed using some kind of CAD software [3,4].

The process of 3D printing has three basic components: computer-assisted (i.e. digital) design, machine equipment, and an added material. Different materials can be used to build 3D-printed objects, from plastic, metal and rubber to human cells. A 3D printer works in similar ways as a regular printer. 3D Printing essentially describes an assortment of technologies that digitally formulate three-dimensional objects on a layer-by-layer basis. It has been adopted by students, entrepreneurs, hobbyists, and various industries. As shown in Figure 8.1, 3D printing involves three steps [5]. A typical 3D printer is shown in Figure 8.2 [6]. The business uses of 3D printing are growing year by year.

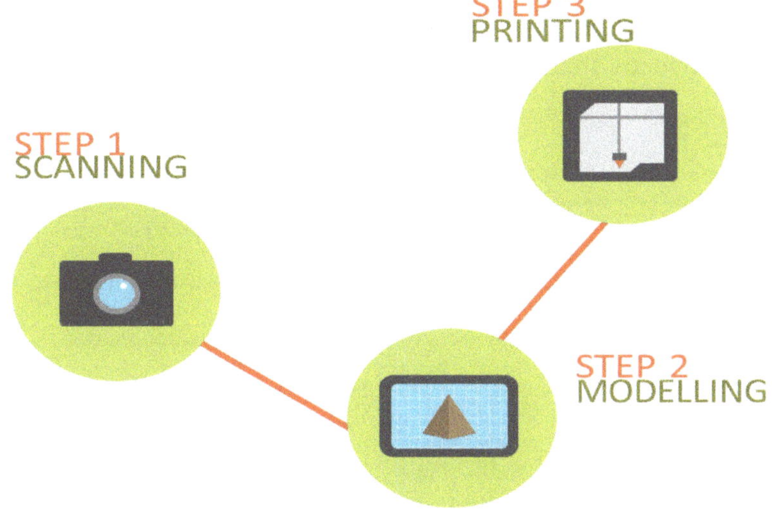

Figure 8.1 3D printing involves three steps [5].

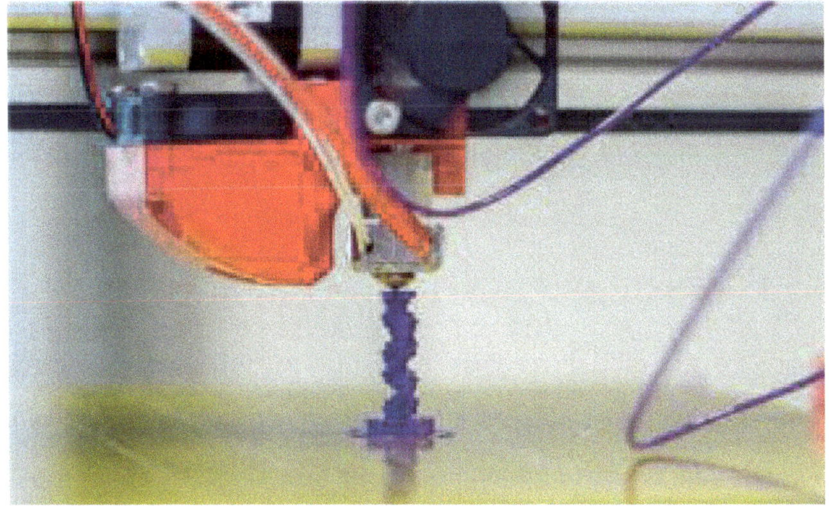

Figure 8.2 A typical 3D printer [5].

In essence, 3D printing is a manufacturing process in which material is laid down, layer by layer, to form a three-dimensional object. 3D printing can create physical objects from a geometrical representation by successive addition of material. 3D printing is an umbrella concept for a set of processes and technologies that offer a wide range of the production of parts and products in different materials. Varieties of 3D printing technologies have been developed with different functions. One thing common in all these processes is the manner in which production is carried out – layer by layer in an additive process. Not all 3D printers use the same technology.

Depending on the technology, 3D printers can use a variety of materials, including but not limited to metals (stainless steel, solder, aluminum, and titanium among them); plastics and polymers (including composites that combine plastics with metals, wood, and other materials); ceramics; plaster; glass; and even foodstuffs like cheese, icing, and chocolate. Even composites can be used. The choice of material greatly depends on the product's function and desired properties [7,8].

8.3 TYPES OF 3D PRINTING

There are generally three types of additive manufacturing: selective binding, selective solidification, and selective deposition. Typically, people refer to these technologies as Selective Laser Sintering

(SLS), Stereolithographic (SLA), and Fused Deposition Modeling (FDM), which are discussed as follows [9,10].

- *Stereolithography* (SLA): This was the world's first 3D printing technology, invented in the 1980s. It is an additive manufacturing process that employs a vat of liquid ultraviolet curable photopolymer "resin" and an ultraviolet laser to build parts' layers one at a time. For each layer, the laser beam traces a cross-section of the part pattern on the surface of the liquid resin. SLA parts have the highest resolution and accuracy and the smoothest surface finish of all plastic 3D printing technologies. Although stereolithography can produce a wide variety of shapes, it has often been expensive.

- *Selective laser sintering* (SLS): This is an additive manufacturing technique that uses a high-power laser (for example, a carbon dioxide laser) to fuse small particles of plastic, metal (ceramic, or glass powders into a mass that has a three-dimensional shape. The SLS machine preheats the bulk powder material in the powder bed somewhat below its melting point. SLS is trusted by engineers and manufacturers across different industries for its ability to produce strong, functional parts. Low cost per part, high productivity, and established materials make the technology ideal for custom manufacturing. The material selection for SLS is limited compared to FDM and SLA.

- *Fused Deposition Modeling* (FDM): This is the most widely used form of 3D printing at the consumer level, fueled by the interest of hobbyists in 3D printers. This technique is suited for basic proof-of-concept models, as well as quick and low-cost prototyping of simple parts. FDM is regarded as a very clean technology, usually simple and office-friendly. It uses a continuous filament of a thermoplastic material. The technology can produce complex geometries and cavities that would otherwise be quite problematic. Since 2004, FDM technology has been used in a particular sector to produce load-bearing scaffold. Home printers based on FDM typically work with plastic filament.

Which technology makes the most sense for you to use depends on your budget, the model's complexity, and the finest detail that is necessary.

8.4 3D PRINTING IN OIL AND GAS

The oil and gas (O&G) industry is among the most demanding and complex domains. Its complexity is demonstrated in Figure 8.3 [11]. 3D printing is already disrupting supply chains in automotive, aerospace, and consumer products. It is in its infancy in the oil &

gas sector. The oil and gas sector lags behind industries like automotive and aerospace in finding applicable use cases for AM. This is possibly due to the widespread adherence to a conservative culture that supports tried and tested methods of production and operation often referred to as "the race to be second." There is considerable resistance to change. Initially, 3D printing was largely limited to polymer-based products, which had limited appeal to the oil and gas industry. However, advances in metal-based printing are making the technology much more relevant. Figure 8.4 shows the oil and gas industry supply chain with its segments and corresponding activities [12]. The segments are explained as follows:

Figure 8.3 The complexity of oil and gas industry [11].

Figure 8.4 Oil and gas industry segments and corresponding activities [12].

- *Upstream Sector*: This includes the exploration and production segment. It is comprised of the following: exploration, drilling, production, and plug and abandonment.

- *Midstream Sector:* This consists of pipes and transportation methods. It is comprised of the following: compressor and pumping stations, geopolitical issues, and maintenance. Additionally, trucking companies, barge companies, and rail services belong to this segment.

- *Downstream Sector*: This is composed of those relating to the consumers, such as manufacturing, petrochemical refining, retail production distribution, and retail. The upstream and downstream segments are characterized by their heavy investments in infrastructure, equipment, and maintenance.

8.5 APPLICATIONS OF 3D PRINTING IN OIL AND GAS

Engineers can use 3D printers to visualize, design, and validate designs without putting in extra time and resources. They create 3D printing facilities across locations to produce tools or parts whenever the need arises. Some of the most common uses for 3D printing in the oil and gas industry include creating jigs and fixtures, functional prototypes, and display and presentation models. Applications of additive manufacturing in O&G industry include the following [13,14]:

- *Rapid Prototyping:* A key application of 3D printing in the O&G sector is rapid prototyping. Time to market is one of the most critical issues any industry faces. With pressure to create solutions quickly, engineers and designers must make quick, accurate decisions during the concept stage. Rapid prototyping is a key step in design verification. It goes hand-in-hand with 3D printing. 3D printing prototypes allow engineers to produce multiple iterations and change a component design overnight to meet deadlines. Figure 8.5 shows a typical rapid prototyping [15].

Figure 8.5 A typical rapid prototyping [15].

• *End-use Parts:* The oil and gas industry requires parts to meet robust performance and environmental standards. Utilizing 3D printing to fabricate production, end-use parts has become an increasingly mainstream operation in the energy sector. Because 3D printing can create custom, complex parts faster than traditional manufacturing processes, engineers have found the technology to be a perfect solution for low-volume projects.

• *Spare Parts:* One increasingly crucial application of 3D printing in the energy sector is in the spare parts market. The high cost of downtime and logistical challenges of distribution to widespread, remote locations have amounted to overstocking of spare parts. 3D printing provides a solution through fast, on-demand printing of legacy parts from an on-site system. 3D printing enables on-site manufacturing, for fast replacement of broken parts. Printing replacement parts quickly and on-site eliminates costly wait times associated with shipping parts from industrial centers located far away. Solving outdated parts is made easy with 3D printing technology. Additive manufacturing requires only enough storage space to keep the part's digital file on hand. It works best when there are no specs available for legacy parts needed for oil drilling rigs that may have been around for decades. In the oil and gas sector, 3D printing is primarily used for rapid prototyping, creating custom parts like gas turbine nozzles, downhole tools, valves, and other components. Figure 8.6 shows a 3D-printed fuel nozzle made by GE engineers [16].

Figure 8.6 3D-printed fuel nozzle made by GE engineers [16].

- *Metal Parts:* Metal 3D printing is a process of constructing three-dimensional parts from metal powder. It works like the regular 3D printing process. The main benefit of metal 3D printing is its ability to create complex parts without needing multiple steps or processes. Metal 3D printing also allows for more efficient use of materials since only what is needed is used, so waste can be minimized based on design specifications. Any metal that can be ground into powder and melts at a high temperature can, in theory, be utilized.

- *Reverse Engineering:* 3D printing can make reverse engineering custom or legacy parts for oil drilling rigs more efficient than ever before. By using digital designs and 3D scanning, a detailed model and exact measurements of a part can be created from scratch. Reverse engineering new parts through 3D printing reduces downtime caused by equipment breakdowns. It also leads to optimization, customization, and improvements for your machinery.

8.6 BENEFITS

3D printing technology has become more efficient, with lightweight components, cost-efficient services and environmentally friendly materials. The entire supply chain for the maintenance and operation of drilling rigs has been completely transformed by additive manufacturing. 3D printing offers significant benefits to the oil and gas industry by enabling the creation of complex, customized parts on demand, leading to improved design flexibility, faster production times, cost reductions, and enhanced operational efficiency. Even the most demanding components for a rig or plant can be produced locally, cutting down significantly on transportation costs and customs delays. The attractive benefits of 3D printing are numerous. These include the following [14,16]:

- *Prototyping:* Rapid prototyping and bespoke engineering support are recognized as key benefits of 3D printing. 3D printing provides rapid and accurate generation of prototypes and components, specifically where original molding casts no longer exist. 3D printing allows for quick and inexpensive prototyping. It can create custom parts faster than traditional manufacturing methods. AM supports innovation by enabling rapid iteration of physical objects during R&D.

- *Operating in Remote Areas:* Although the oil and gas industry is a crucial economic sector, it faces many challenges from operating in remote areas. Sourcing spare parts can be a challenge. Manufacturing often occurs thousands of miles away from drilling sites. But with the introduction of 3D printing technology, these problems are becoming easier to address.

- *On-site Manufacturing:* 3D printing allows companies to produce parts on-site, which can reduce downtime and costs. This is especially useful in remote locations where shipping parts can be expensive and time-consuming. 3D printing can support the local economy by retaining production within the country.

- *Customization:* The 3D printing technology allows for customization on smaller production runs since each component can be tweaked during manufacture through software manipulation, ensuring accuracy during production without loss of quality or performance.

- *Parts On-demand:* 3D printing can be used to quickly print spare parts on demand. By allowing companies to produce parts on demand, 3D printing can reduce lead times significantly and virtually eliminate the need for storage space. This means that when an operator needs a part quickly, they no longer have to wait weeks or months for the part to be delivered. Instead, they can simply print it themselves onsite with an on-site 3D printer. Additive manufacturing allows engineers to create parts on demand with minimal setup time.

- *Cost Reduction:* AM eliminates the cost of capital associated with stocking inventory that may never get used—because only digital files are needed to print parts. In other words, 3D printing offers immense cost savings and efficiency gains throughout the entire supply chain in the oil and gas industry by reducing lead times and storage costs. The industrial-scale additive manufacturing capabilities have enabled O&G companies to not only drastically reduce costs but also save labor and capital.

- *Reducing Carbon Footprints:* AM could help the oil & gas, offshore and maritime industries decarbonize operations in the energy transition. Distributed just-in-time production close to or exactly where products are required reduces transportation needs. This in turn means less exhaust emissions, including greenhouse gases.

- *Less Waste*: 3D printing generally produces less waste than conventional methods that usually involve machining away material from a large piece of material. In contrast, 3D printing adds only material that is needed, and co-locating it with the recycling of materials could boost local circular economies.

- *Speed*: Speed is important for most oil and gas projects. One of the key benefits of 3D printing is its ability to speed up the product development process. Speed of delivery from order receipt to manufacture to avoid lengthy procurement processes is regarded as a key advantage. This should be compared against traditional methods of order placement, and delivery.

• *Design Freedom:* 3D printing allows for more design freedom, which can lead to higher-quality parts with unusual geometries. In addition to speed and efficiency, metal 3D printing provides design freedom that is not available in traditional manufacturing methods due to its unique layered approach.

• *Design Optimization:* AM allows for intricate geometries and internal structures that can be tailored to specific applications, resulting in lighter, stronger, and more efficient components compared to traditional manufacturing methods.

• *Material Flexibility:* AM can utilize a wide variety of materials, including high-performance metals and polymers, allowing for customized part properties to suit demanding operating conditions.

• *Complex Geometries:* Printing can produce complex components that may be impossible to manufacture using conventional processes. AM can create parts with intricate features, like internal cooling channels or complex interlocking components, which may be impossible with traditional manufacturing techniques.

8.7 CHALLENGES

With any new technology, the additive manufacturing in O&G sector faces some challenges. Potential challenges include batch inconsistency of materials for AM, variation in hardware, differing environmental and operational conditions, and regulatory change. Other challenges include the following [14,17]:

- *Toxicity:* Metal additive manufacturing processes release toxic VOCs (volatile organic compounds), which can have serious health consequences. 3D printing metal parts present a number of potential workplace risks that must be taken seriously by all involved workers and employers alike.

- *Safety:* Safety is the most critical issue in the oil and gas industry, where even minor equipment failures can have catastrophic consequences. When it comes to 3D printing, the safety and reliability of the processes and products is paramount. When printing metal parts using a 3D printer, there are several potential dangers that can arise due to the use of metal powder and laser-based processes. Small particles of metal powder can become airborne during the 3D printing process, meaning they can be accidentally inhaled. All safety protocols must be followed rigorously in order to ensure that everyone remains safe while working with this technology. It is absolutely necessary that workers wear proper personal protective equipment to protect themselves from potential exposure. Strict guidelines should be put in place to prevent any eating or drinking in the printing environment, as this could lead to contamination.

• *Standardization*: The absence of standardization and certification is often cited as one of the key challenges that affect the wider adoption of 3D printing technologies in the oil and gas sector. AM needs a systematic qualification process, standards, and specialized knowledge towards qualification and certification of 3D-printed components and materials. Without standardization, printed parts and components in oil & gas industries could raise the risk of unexpected or premature failures due to inherent variations of mechanical and metallurgical properties associated with the AM parts. Non-standard practices for testing parts raise the probability of overall material costs rising compared to the traditional manufacturing route.

• *Trust*: Greater trust and understanding in the industry of the benefits of 3D printing is required if adoption is to reach its potential growth.

• *Lack of Awareness:* The lack of knowledge of 3D printing, particularly at senior levels in the oil and gas industry, is a challenge. There is a lot of ignorance in the oil and gas industry around new technology – people just do not understand the possibilities. The lack of awareness of 3D printing is concerning and must be addressed if the oil and gas industry is to make the most of the range of technology benefits available to it.

- *Reducing Carbon Emissions:* Oils and gas companies account for 33% of direct and indirect carbon emissions in the world. Many oil and gas companies reduce carbon emissions by switching from subtractive manufacturing to additive manufacturing. An oil and gas company can significantly reduce carbon emissions using 3D printing instead of subtractive manufacturing.

- *Quality Control*: Implementing robust quality control measures to guarantee the integrity of AM parts is essential. For this reason, it is vital that companies consider ways to ensure production repeatability to improve confidence in 3D-printed parts and simplify the path to certification.

- *Intellectual Property Rights (IPR):* 3D printing is facing a strong challenge of IPR. This challenge is also faced by sectors like automotive and aerospace. This challenge will also be faced by the oil & gas industry.

8.8 CONCLUSION

It is evident that 3D printing has opened up new possibilities for the O&G industry. It not only helps reduce downtime through quick repairs but also reduces costs associated with sourcing replacement parts from remote locations. Major companies across the oil and gas industry have adopted industrial-scale additive manufacturing capabilities for production. The market growth potential for 3D printing in the oil and gas industry is significant and growing. Additive manufacturing has the potential to significantly transform the oil and gas industry by enabling the design and production of highly customized, efficient components, leading to improved operational performance and cost savings across the entire supply chain. More information about additive manufacturing in oil & gas industry can be found in the books in [18,19] and the following related journals/magazines:

- *International Journal Of Petroleum And Petrochemical Engineering*

- *Journal Of Petroleum Science And Engineering*

- *Petroleum*

- *Petroleum Research*

- *Energy Reports*

- *Oil & Gas Journal*

REFERENCES

[1] P. Madeleine, "What are the 3D printing applications in the oil & gas sector?" September 2022,

https://www.3dnatives.com/en/3d-printing-applications-in-the-oil-gas-sector-220920225/

[2] M. N. O. Sadiku, M. Oteniya, J. O. Sadiku, and S. Abunene, "3D printing in oil and gas industry," *International Journal of Trend in Scientific Research and Development*, vol. 8, no. 5, September-October 2024, pp. 853-860.

[3] M. N. O. Sadiku, U. C. Chukwu. A. Ajayi-Majebi, and S. M. Musa, "3D printing: An introduction, " *International Journal of Trend in Scientific Research and Development,* vol. 6, no. 7, November-December 2022, pp. 573-577.

[4] F. R. Ishengoma and T. A. B. Mtaho, "3D printing: Developing countries perspectives computer engineering and applications," *International Journal of Computer Applications*, vol. 104, no. 11, October 2014, pp. 30-34.

[5] D. Pitukcharoen, "3D printing booklet for beginners,"

https://www.metmuseum.org/-/media/files/blogs/digital-media/3dprintingbookletforbeginners.pdf

[6] S. Torta and J. Torta, *3D Printing: An Introduction*. Mercury Learning and Information, 2019.

[7] T. Hoffman, "3D printing: What you need to know," July 2020,

https://www.pcmag.com/news/3d-printing-what-you-need-to-know

[8] "Beginner's guide to 3D printing,"

https://www.think3d.in/landing-pages/beginners-guide-to-3d-printing.pdf

[9] "Introduction to 3D printing,"

https://education.gov.mt/en/resources/news/documents/youth%20guarantee/3d%20printing.pdf

[10] "Guide to 3D printing materials: Types, applications, and properties,"

https://formlabs.com/blog/3d-printing-materials/

[11] "Is the oil & gas industry ready for 3D printing?" September 2021,

https://amfg.ai/2021/09/06/is-the-oil-gas-industry-ready-for-3d-printing/#:~:text=While%20the%20oil%20and%20gas,unsuitable%20for%20low%2Dvolume%20production.

[12] E. B. Caldona et al., "Additively manufactured high-performance polymeric materials and their potential use in the oil and gas industry," *Chemistry and Materials for Hydrocarbon Recovery Prospective*, vol. 11, December 2021, pp. 701–715.

[13] "3D printing in energy," July 2018

https://www.stratasys.com/en/stratasysdirect/resources/articles/energy-oil-gas/

[14] "How 3D printing is changing the oil & gas industry,"

https://www.arnolditkin.com/blog/oilfield-accidents/how-3d-printing-is-changing-the-oil-gas-industry/#:~:text=3D%20printing%20can%20make%20reverse,can%20be%20created%20from%20scratch.

[15] "3D printing,"

https://www.shell.com/what-we-do/digitalisation/3d-printing.html

[16] "3D printing in oil & gas industry – Advantages & challenges," January 2019,

https://manufactur3dmag.com/3d-printing-in-oil-gas-industry-advantages-challenges/

[17] T. Pantelis, "Designing the future: The benefits of additive manufacturing to oil and gas,"

https://www.hazardexonthenet.net/article/192101/Designing-the-future--the-benefits-of-additive-manufacturing-to-oil-and-gas.aspx

[18] G. Blokdyk, *3D Printing in Oil and Gas.* 5STARCooks, 2nd edition, 2021.

[19] A. Tarancón and V. Esposito (eds.), *3D Printing for Energy Applications.* Wiley, 2021.

CHAPTER 9

DRONES IN OIL AND GAS

"I think the greatest threat to the privacy of Americans is the drone, and the use of the drone and the very few regulations that are on it today, and the booming industry of commercial drones."

– Dianne Feinstein

9.1 INTRODUCTION

Drones have proven themselves a unique and powerful asset through all their work. As a result, all industries are using them today to get various tasks done. Industries like e-commerce, defense, entertainment, manufacturing, telecommunications, etc. have been using drones since their early development. The application of drone technology has transformed many industries, including the oil and gas industry. Drones have become a crucial part of the oil and gas industry, where they are being deployed to perform a wide variety of tasks. Drones in oil and gas have helped reduce inspection time, cut costs, decrease downtime, and identify problems early on. When it comes to oil and gas industry, drones are a super significant investment. They have made it possible to automate inspections, ease management, and save lives [1]

The oil and gas industry, known for its vast operations and extensive infrastructure, is undergoing a technological renaissance with the integration of drone inspections. An example of such extensive infrastructure is shown in Figure 9.1 [2]. The oil and gas sector has long been at the forefront of adopting cutting-edge technologies to enhance safety, efficiency, and cost-effectiveness. Drones are transforming the oil and gas industry, offering a range of applications that enhance operational efficiency, safety, and environmental compliance. By leveraging the capabilities of drone technology, oil and gas companies can streamline operations, reduce costs, improve safety outcomes, and make informed decisions based on real-time data.

Figure 9.1 An extensive infrastructure for oil and gas plant [2].

Drones are transforming oil and gas exploration and providing numerous benefits for the companies that use them. They enable companies to optimize their operations, improve safety, and minimize environmental risks, making them indispensable in the oil and gas industry. They offer a cost-effective solution for routine inspections and data collection, reducing the need for extensive manpower and helicopter-based surveillance. One driving factor behind the remarkable oil and gas drone market growth is the cost-effectiveness and efficiency they bring to the industry. Consequently, cost-conscious companies increasingly turn to drones to optimize their operations, making this a pivotal driver of market expansion [3].

This chapter examines the applications of drones in the oil and gas industry. It begins with explaining what drones are all about. It covers drones in the oil and gas industry. It presents some applications of drones in oil and gas. It highlights the benefits and challenges of drones in oil and gas. The last section concludes with comments.

9.2 WHAT IS A DRONE?

The FAA defines drones, also known as unmanned aerial vehicles (UAVs), as any aircraft system without a flight crew onboard. Drones include flying, floating, and other devices, including unmanned aerial vehicles (UAVs), that can fly independently along set routes using an onboard computer or follow commands transmitted remotely by a pilot on the ground [4]. A typical drone is shown in Figure 9.2 [5]. A drone is usually controlled remotely by a human pilot on the ground, as typically shown in Figure 9.3 [6]. Drones can range in size from large military drones to smaller drones. Drones, previously used for military purposes, have started to be used for civilian purposes since the 2000s. Since then, drones have continued to be used in intelligence, aerial surveillance, search and rescue, reconnaissance, and offensive missions as part of the military Internet of things (IoT). Today, drones are used for different purposes such as aerial photography, surveillance, agriculture, entertainment, healthcare, transportation, law enforcement, etc.

Figure 9.2 A typical drone [5].

Figure 9.3 A drone is usually controlled by operators on the ground [6].

Commercial drones have come a long way in the last decade. Drones work much like other modes of air transportation, such as helicopters and airplanes. When the engine is turned on, it starts up, and the propellers rotate to enable flight. The motors spin the propellers and the propellers push against the air molecules downward, which pulls the drone upwards. Once the drone is flying, it is able to move forward, back, left, and right by spinning each of the propellers at a different speed. Then, the pilot uses the remote control to direct its flight from the ground [7].

Drone laws exist to ensure a high level of safety in the skies, especially near sensitive areas like airports. They also aim to address privacy concerns that arise when camera drones fly in residential areas. These include the requirement to keep your drone within sight at all times when airborne. In the United States, drones weighing less than 250g are exempt from registration with civil aviation authorities. If your drone exceeds 250g in weight, you will also require a Flyer ID, which requires passing a test [8]. It is necessary to register as an operator, be trained as a pilot, and have civil liability insurance, in addition to complying with various flight regulations, and those of the places where their use is permitted.

Most drones have a limited payload, usually under 11 pounds. Drones are classified according to their size. Here are the different drone types:

- Nano Drone: 80-100 mm

- Micro Drone: 100-150 mm

- Small Drone: 150-250 mm

- Medium Drone: 250-400 mm

- Large Drone: 400+ mm

One of the emerging trends in drone use for factories is the utilization of LiDAR technology. LiDAR stands for Light Detection and Ranging. This technology provides accurate depth information essential for understanding the three-dimensional structure of the environment. LiDAR sensors emit laser beams to measure distances to objects, creating high-resolution 3D maps of the surrounding terrain and objects. The ability to capture detailed data through LiDAR technology has opened up opportunities for better predictive maintenance, reduction in inspection times, and overall cost savings [9].

9.3 OIL & GAS DRONES

The oil and gas industry is one of the biggest industries in the world. It is one of the most capital-intensive while being one of the largest contributors to the global economy. Oil and gas firms are regarded as one of the critical industries. The oil and gas industry has traditionally been a conservative sector, hesitant to embrace new technologies. However, the advent of drones has provided the industry with new opportunities to optimize operations, reduce costs, and improve safety. In the last decade, oil and gas companies have harnessed the mobility and perspective of drones to innovate workflows and save money.

North America, particularly the US, leads the world in drone adoption, while European markets follow closely. In 2013, BP became the first oil and gas company to receive a license to operate drones. Several oil and gas companies such as ExxonMobil, Shell, and Chevron followed suit. The oil and gas drone market size is experiencing robust growth due to several key factors. There is a growing emphasis on cost reduction and operational efficiency in the energy sector.

There are drones specifically designed for the oil and gas industry. Oil and gas drones encompass a range of specialized UAV technologies tailored for the oil and gas industry. These drones are equipped with high-resolution cameras, sensors, and data analysis software. The drones provide crucial data for inspections, maintenance, and exploration. The drones can be used to monitor pipelines, check for leaks, and conduct surveys of oil spill zones. They can also fly over rough terrain, making them ideal for inspections of pipelines and other infrastructure placed in unmanageable terrain. The drone technology allows for increased efficiency, improved safety, access to remote areas, cost-effectiveness, enhanced data collection, and improved response to accidents [10]. A growing variety of sensors allow oil and gas drones to perform more tasks and make these tasks more sophisticated. Today, the oil and gas industry cannot do without drones. A typical oil and gas drone is displayed in Figure 9.4 [6].

Figure 9.4 A typical oil and gas drone [6].

9.4 APPLICATIONS

Drone deployment has significantly transformed the oil and gas industry, especially for operations in difficult-to-reach areas such as offshore rigs and pipelines. Drones have been applied in diverse capacities ranging from drone inspection, asset inspection, leak detection, pipeline inspection, flare stack inspection, and emergency response, etc. As more and more commercial drone manufacturers work closely with oil and gas companies around the world, more customized applications are made available. Advancements in drone technology are also increasing its applications. Common applications of drones in oil and gas sector include the following:

- *Inspection:* Oil & gas inspection applications are at the forefront of drone adoption in the industry. Drones can collect better data than humans can because they can get much closer to infrastructure. The drones are deployed for critical tasks such as pipeline monitoring, rig inspections, and facility surveillance, enhancing safety, and operational efficiency. While environmental impact assessment and other applications are significant, it is the pressing need for infrastructure inspection and maintenance that drives the dominance of the oil and gas inspection segment. Drones can run inspections without needing to shut down oil operations. They help you save on several maintenance costs, which in turn frees up funds to conduct more frequent drone inspections. An example of a drone for oil and gas inspection is shown in Figure 9.5 [11].

Figure 9.5 A drone for oil and gas inspection [11].

- *Flare Stack Inspection*: Flare stacks, used to dispose of waste gases safely, require regular inspections to ensure their integrity and functionality. Traditionally, manual inspections of flare stacks involve significant downtime and safety risks for workers. Drones can capture detailed images, assess heat signatures, and detect potential issues, enabling proactive maintenance and minimizing downtime. Figure 9.6 shows flare stacks in operation [12].

Figure 9.6 Flare stacks in operation [12].

• *Leak Detection:* Before drones, oil and gas companies tried to detect leaks by mounting fixed detectors at high-risk spots in facilities and along pipelines or by having inspectors occasionally check areas with portable detectors. However, these traditional leak detection methods can be costly and inefficient. Drones equipped with gas detectors can quickly identify the location of leaks and help determine the extent of the leak. Drone-based leak detection is faster and more comprehensive. Drones can easily detect spills and corrosion in flare stacks that are high off the ground. Drones equipped with sensors and artificial intelligence can detect gas leaks, pipeline corrosion, and crude oil spills, enabling emergency response teams to take action quickly.

- *Environmental Monitoring:* Drones can now be equipped with multispectral and hyperspectral cameras and are used to monitor the environmental impact of oil and gas operations. They can assess soil health, vegetation health, and water quality in and around drilling sites. This way, companies can adhere to environmental regulations, ensure responsible resource extraction, and minimize ecological disruption.

- *Security:* Drones add an extra layer to field security by patrolling expansive oil fields, pipelines, and remote facilities. Drones can patrol remote facilities, pipelines, and oil fields to detect unauthorized activities, trespassing, and security breaches. They provide real-time visual feeds to security personnel, enabling rapid responses and enhancing overall site security. They can be really helpful in identifying and mitigating safety risks. They can be a great tool for managing risks by reducing human exposure to hazardous environments, enhancing worker safety, and supporting adherence to strict safety protocols.

- *Data Collection:* Drones can provide high-quality data in a fraction of the time and at a lower risk to humans. Drones can be equipped with advanced sensors to gather massive amounts of data during flyovers. This data is then processed and analyzed using specialized software. The result guides reservoir management decisions, optimizes drilling techniques, and improves overall production efficiency. The incorporation of AI and data analytics will make data collection and analysis quicker, more accurate, and more comprehensive.

- *Monitoring:* The need for autonomous monitoring and inspection in the oil and gas industry is clear. Oil and gas companies use drones to remotely check and observe equipment, infrastructure components, trucks, tankers, and other company assets. Drones help monitor and manage long pipelines spread across different geographies, identifying leakages or anomalies quickly. The drones are able to deliver 360-degree views of subjects for the monitoring of field operations, keep an eye on the development of new facilities, and detect encroachment on pipelines, railways, and other valuable company property. Drone-powered inspections and safety monitoring improve efficiency without halting operations or compromising personnel safety.

9.5 BENEFITS

The use of drones in oil and gas industry has several benefits that make exploration safer, more efficient, and cost-effective. Drones can save millions of dollars in asset maintenance costs. They also allow you to cut down on labor costs. Their combination of safety, efficiency, precision, and cost-effectiveness makes them an indispensable tool in the modern oil and gas inspection toolkit. Other benefits of drones in oil and gas industry include the following [2,13]:

- *Safer & Faster Inspections:* Traditional oil inspections are dangerous, to say the least. But drones are making inspections much safer. These traditional inspection methods can be costly and inefficient. Drones inspect better because of their advanced cameras. They can fly to great heights and through toxic chemicals with ease, so there is no need to put personnel in harm's way. While traditional oil and gas inspections can take several days to complete, drones fly directly to a target at command, perform comprehensive checks, and eliminate the need to shut down operations. Nearly every stage of petroleum production benefits from close-eyed inspections by drones.

- *Increased Efficiency:* Drones can collect data at a much faster pace than traditional methods. They can perform routine inspections faster and provide better visibility of assets. This enables oil and gas companies to survey large areas quickly and allows them to detect faults and geographical features at a much faster pace. All of this can be done in a short amount of time and at a reduced cost because drones do not need as much manpower as traditional methods require. Drones have the capacity to reduce operational expenses significantly by eliminating the need for extensive human labor and expensive helicopter-based surveillance.

- *Improved Safety:* Safety is a significant concern in all aspects of the oil and gas sector. In fact, a significant benefit of drone technology is safety. Traditional methods of oil and gas exploration often involve on-site staff who are at risk of physical harm and exposure to hazardous substances. Such on-site workers are shown in Figure 9.7 [14]. By using drones to conduct surveys and inspections, oil and gas companies can avoid putting their staff in danger. Drones can eliminate traditional methods of inspecting areas like essential production components in oil refineries, chimneys, smokestacks, storage tanks, jetties, and other potentially hazardous environments.

Figure 9.7 On-site workers [14].

• *Access to Remote Areas:* Some oil and gas exploration takes place in remote and inaccessible areas. By using drones, oil and gas companies can survey these areas without putting their staff in danger.

• *Cost-Effectiveness:* Drone technology is more cost-effective than traditional methods of oil and gas exploration. By using drones, oil and gas companies can reduce their expenses while still collecting valuable data that can inform their operations. The cost savings extend to safety enhancements, as drones can access hazardous or remote locations without endangering human lives. Drones can be used to inspect thousands of miles of pipelines at a fraction of the cost and at a more precise level.

• *Enhanced Data Collection:* Drone technology is transforming data collection and analysis in the oil and gas industry, allowing greater accuracy and speed. Drones are equipped with sensors that can measure heat, pressure, and humidity levels, which provide real-time data about natural gas and oil reserves.

• *Improved Response to Accidents*: In the event of an accident in the oil and gas industry, time is of the essence. Drones can help oil and gas companies respond more quickly to accidents and provide more accurate information about the extent of the damage.

• *Environmental Regulation:* Stringent safety and environmental regulations are pushing oil and gas companies to adopt advanced technologies for monitoring and maintaining their infrastructure. Stricter environmental regulations and heightened concerns over the ecological impact of the industry have pushed energy companies to adopt more responsible practices.

• *Industry Giants:* The oil and gas drone market is significantly influenced by key industry giants that play a pivotal role in driving market dynamics and shaping consumer preferences. Their strong global presence and brand recognition have contributed to increased consumer trust and loyalty, driving product adoption. These industry giants continually invest in research and development, introducing innovative designs, materials, and smart features.

• *Emergency Response:* One of the key challenges in emergency situations is getting accurate information to the decision-makers in real-time. It is important to have a strategy in place when disasters happen. Disasters may be oil spills, natural, mechanical blowouts, etc. Drones can help you respond quickly and efficiently. Unlike manned aircraft that need a pilot and lead time to be ready for takeoff, drones can launch immediately. Their speedy deployment can save precious time when disaster hits. Drones can also ensure emergency response equipment is set up in the right spots.

• *Deliveries:* Drones can carry out deliveries of material, replacement parts, and supplies to offshore rigs and other remote operations, instead of sending out a boat or helicopter to reach the rig.

• *Real-Time Communication:* Communication is crucial during exploration, inspection, production, and other processes, as there should be clear and prompt data transmission between professionals and departments involved. Proceeding with oil and gas drone inspection also secures real-time communication. The drone pilot can capture and record the situation and directly send it to authorized offices for viewing and analysis.

•   *Accurate Data Gathering:* Manual oil and gas topside corrosion diagnoses and inspections are manual activities prone to many bottlenecks, and prone to inaccuracy. What makes drone technology an invention worth the investment is that it secures accurate data that can help in the success of the work operation. Drone manufacturers are continuously improving their products by meeting the needs of different industries. The best thing to expect with drones is the ability to record and present precise and detailed imaging data that can help track and analyze the work process.

•   *Environmental Impact:* The oil and gas sector is under increasing pressure to minimize environmental impact. Drones offer a more eco-friendly alternative to conventional methods. They operate silently, minimize ground disturbances, and leave a reduced carbon footprint. Their ability to operate without disturbing local ecosystems makes them an environmentally responsible choice.

•   *Accessibility:* There are areas within the vast landscapes of the oil and gas sector that are incredibly hard to reach, or even inaccessible through traditional means. Drones effortlessly navigate these areas, ensuring that no asset remains unchecked.

• *Continuous Monitoring:* Drones enable 24/7 monitoring of assets and infrastructure. Drones can be programmed to conduct scheduled or on-demand inspections, capturing high-resolution images and videos for analysis.

9.6 CHALLENGES

The oil and gas industry faces numerous challenges, such as remote and harsh operating environments, high safety risks, the need for cost-effective solutions, and the inherent risks associated with operations. Other challenges of drones in oil and gas industry include the following [11]:

• *Complex Regulation:* One significant challenge facing oil and gas drones is the complex regulatory environment. Drones operate in an airspace that is governed by a myriad of regulations. In many regions, there are restrictions on flight altitudes, no-fly zones, and specific permissions required for commercial use. Companies must adhere to various national and international regulations concerning drone usage, such as registration, flight restrictions, and operator certifications. The use of drones in the energy sector is subject to strict aviation regulations, which can vary by region and country. Ensuring drones adhere to both aviation and industry-specific guidelines adds complexity to their deployment. Overcoming these regulatory and compliance hurdles is a challenge for market players, which can impede the seamless integration of drones into oil and gas operations. There is good news for companies that build drone applications for oil and gas: the regulatory climate seems to be finally starting to relax.

- *Safety:* One of the paramount concerns in the oil and gas sector is ensuring the safety of its personnel. Employee safety is of paramount importance in refineries, which are inherently hazardous environments. There are safety concerns such as drones entering restricted airspace, collisions with other aircraft, and human error that can impact operations. With emissions regulation tightening and public awareness of environmental issues growing, the oil and gas industry is under increasing pressure to take safety seriously. Drones eliminate direct human exposures, allowing for inspections in even the most challenging and hazardous environments without jeopardizing human safety.

- *Privacy:* Concerns related to privacy and data security further complicate their widespread adoption.

- *Cost:* Purchasing drones can be expensive, considering that they will also require a considerable investment in terms of drone pilots, training, and maintenance.

- *Expertise Required:* Drone operation and analysis require specific skills and expertise that can be a challenge for the oil and gas industry. Operating a drone, especially in complex environments like oil and gas facilities, requires specialized skills. Companies need to train drone pilots who can operate the vehicles safely, follow flight plans, and analyze the data collected. Additionally, they need data scientists who can analyze the information and provide insights that are actionable.

- *Technical Limitations:* In spite of their advanced capabilities, drones do have limitations. The range of drones can be limited, implying that they can only operate in a specific area, limiting the extent to which they can be used to replace manual inspection and monitoring activities. Battery life can restrict the duration of inspection flights, which is reduced by cold temperatures. Drones also face challenges operating in extremely windy and rainy weather conditions, limiting their operational days.

- *Data Management:* In the vast world of oil and gas inspections, drones have emerged as technological marvels, capturing massive amounts of detailed data. A fundamental challenge with drone inspections is the sheer volume of data collected. Drones can capture massive amounts of data in a single flight. While this is an advantage, it can also lead to information overload that needs effective storage and analysis. Without an efficient system or platform to manage this data, crucial insights might be overlooked, or the data may simply be inaccessible to those who need it.

- *Integration with Existing Systems:* Companies in the oil and gas sector often have legacy systems in place. Integrating drone-captured data with these existing systems can be challenging and might require additional software or platform solutions.

- *Environmental Concerns:* While drones are relatively less invasive, they can sometimes raise concerns related to noise, privacy, and potential disturbances to local wildlife. Ensuring that drone operations are environmentally sensitive and respectful of local communities is essential.

9.7 CONCLUSION

Specialists in many industries are exploiting the unique flexibility and observational capabilities of drones to improve industrial processes and operational efficiency. Drones have come into use in the oil and gas industry as a precise, highly maneuverable, and cost-efficient means of carrying out transportation network and asset inspections. The use of drones has significantly reduced the risk to personnel and improved the efficiency of inspections.

The adoption of drone technology will become increasingly common throughout the oil and gas industry as companies discover the advantages of using drones. As the industry moves forward, it is evident that the synergy between drone technology and advanced data management platforms will define the future of oil and gas inspections. Oil and gas companies that embrace drone technology will have a significant advantage. One major trend in the oil and gas drones is the integration of artificial intelligence (AI) and machine learning. Drones equipped with advanced AI algorithms can autonomously detect anomalies in infrastructure, predict maintenance needs, and optimize operational processes. The future of drones in the oil and gas industry looks bright, with advancements in technology continually expanding their capabilities.

REFERENCES

[1] G. Goyal, "Why is drone usage growing exponentially in oil and gas industry?" September 2023,

https://www.iotforall.com/why-is-drone-usage-growing-exponentially-in-oil-and-gas-industry#:~:text=Drones%20play%20a%20crucial%20role,or%20corrosion%20along%20pipeline%20surfaces.

[2] "5 Major benefits of drones in oil and gas,"

https://thedronelifenj.com/benefits-of-drones-in-oil-and-gas/

[3] M. N. O. Sadiku, O. D. Olaleye, and J. O. Sadiku, "Drones in oil & gas industry," *International Journal of Trend in Research and Development*, vol. 11, no. 5, September 2024, pp. 96-101.

[4] M. N. O. Sadiku, P. A. Adekunte, and J. O. Sadiku, "A rrimer on drones," *International Journal of Trend in Scientific Research and Development,* vol. 8, no. 4, July-August 2024, pp. 212-220.

[5] "IIow are drones used in thc oil and gas industry?" July 2021,

https://copas.org/how-are-drones-used-in-the-oil-and-gas-industry/

[6] "Oil & gas,"

https://enterprise.dji.com/oil-and-gas

[7] "How drones work and how to fly them," May 2024,

https://dronelaunchacademy.com/resources/how-do-drones-work/

[8] "What are the main applications of drones?" June 2024,

https://www.jouav.com/blog/applications-of-drones.html

[9] "Drones in manufacturing: A game-changer for industry,"

https://viper-drones.com/industries/infrastructure-drone-use/manufacturing/#:~:text=The%20integration%20of%20drones%20into,on%20manufacturing%20is%20no%20exception.

[10] "Oil and gas drones market," August 2024,

https://www.businessresearchinsights.com/market-reports/oil-and-gas-drones-market-109938

[11] "Oil and gas drone inspections: Accelerating asset management," September 2023,

https://optelos.com/oil-and-gas-drone-inspection/

[12] "Revolutionising the oil and gas sector: The rise of drone-in-a-box systems," May 2023,

https://www.linkedin.com/pulse/revolutionising-oil-gas-sector-rise-drone-in-a-box

[13] "6 Unique advantages of drone technology in oil and gas exploration,"

https://www.linkedin.com/pulse/6-unique-advantages-drone-technology-oil-gas-ryan-shore#:~:text=The%20technology%20allows%20for%20increased,and%20improved%20response%20to%20accidents.

[14] "Drones in oil and gas industry," Unknown Source.

CHAPTER 10

CYBERSECURITY IN OIL AND GAS

"Cybercrime is the greatest threat to every company in the world."

– Ginni Rommety

10.1 INTRODUCTION

The oil and gas industry is a prime target for cyber-attacks due to the high value of the data and systems they control. Cyber-attacks on the oil and gas industry are growing because the sector is becoming increasingly dependent on technology and automation. The industry's production, transportation, and refining processes rely heavily on control systems connected to the Internet, making them vulnerable to cyber threats [1]. Oil and gas companies, being critical infrastructure, can become prime targets for state-sponsored or politically motivated cyber groups aiming to disrupt operations, steal sensitive data, or cause economic damage. Gas producers may face a range of security risks, from production shutdowns to inaccessibility. This could result in long-term shortages [2].

Cybersecurity is a critical aspect of the oil and gas industry because it protects the sensitive data and operational technology that the industry relies on. Oil and gas operations involve critical infrastructure such as refineries, pipelines, and drilling rigs. These operations are vulnerable to cyberattacks. The consequences of successful cyberattacks can be severe, leading to physical damage, production disruptions, environmental disasters, and significant financial losses.

Oil and gas (O&G) companies deal with vast amounts of sensitive information, and protecting it is vital to prevent financial loss, reputational damage, and regulatory non-compliance. Playing a vital role in the global economy, the oil and gas industry is a prime cyber threat target. A cyberattack on O&G critical infrastructure could cause physical, environmental, and economic harm and broad disruptions to oil and gas supplies and markets. Cybersecurity is paramount for the oil and gas industry as it plays a critical role in securing the sensitive data and operational technology that enables the industry to extract, produce, and transport oil and gas [3].

The chapter focuses on systematically exploring cybersecurity and safety challenges of the O&G sector. It begins with explaining what cybersecurity is all about. It covers cybersecurity in the oil and gas industry. It describes the roles of cybersecurity in oil and gas. It highlights the benefits and challenges of cybersecurity in oil and gas. The last section concludes with comments.

10.2 OVERVIEW OF CYBERSECURITY

Cybersecurity refers to a set of technologies and practices designed to protect networks and information from damage or unauthorized access. It is vital because governments, companies, and military organizations collect, process, and store a lot of data. As shown in Figure 10.1, cybersecurity involves multiple issues related to people, processes, and technology [4]. Figure 10.2 shows different components of cybersecurity [5].

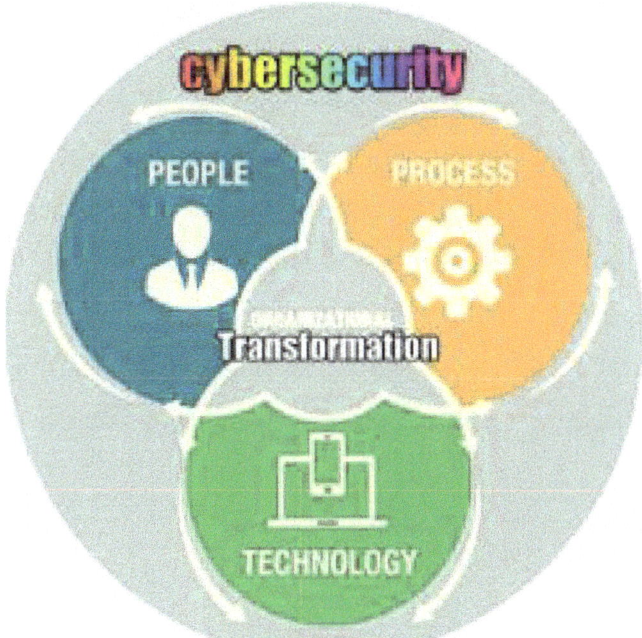

Figure 10.1 Cybersecurity involves multiple issues related to people, process, and technology [4].

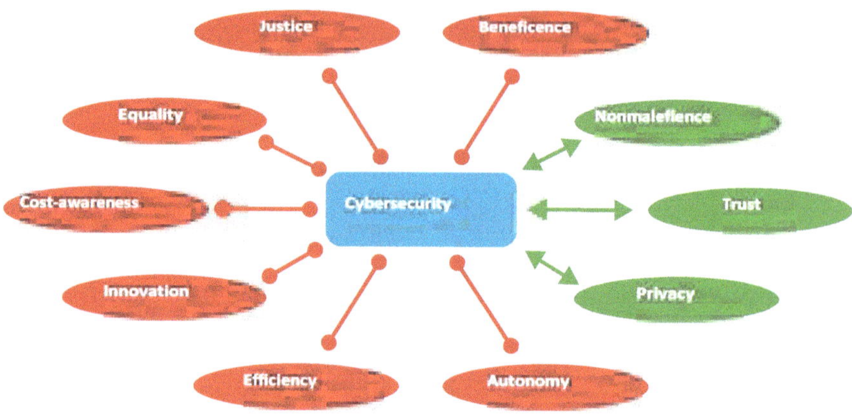

Figure 10.2 Different components of cybersecurity [5].(Green: supportive; red: in tension)

A typical cyber-attack is an attempt by adversaries or cybercriminals to gain access to and modify their target's computer system or network. Cybercriminals or ethical hackers are modern-day digital warriors, possessing extraordinary skills and knowledge to breach even the most impregnable systems. A typical cybercriminal is shown in Figure 10.3 [6]. Cyber-attacks are becoming more frequent, sophisticated, dangerous, and destructive. They are threatening the operation of businesses, banks, companies, and government networks. They vary from illegal crimes of individual citizens (hacking) to actions of groups (terrorists) [7].

Figure 10.3 A typical cybercriminal [6].

The cybersecurity is a dynamic, interdisciplinary field involving information systems, computer science, and criminology. The security objectives have been availability, authentication, confidentiality, nonrepudiation, and integrity. A security incident is an act that threatens the confidentiality, integrity, or availability of information assets and systems [8].

- *Availability*: This refers to the availability of information and ensuring that authorized parties can access the information when needed. Attacks targeting the availability of service generally lead to denial of service.

- *Authenticity*: This ensures that the identity of an individual user or system is the identity claimed. This usually involves using a username and password to validate the identity of the user. It may also take the form of what you have such as a driver's license, an RSA token, or a smart card.

- *Integrity*: Data integrity means information is authentic and complete. This assures that data, devices, and processes are free from tampering. Data should be free from injection, deletion, or corruption. When integrity is targeted, nonrepudiation is also affected.

- *Confidentiality*: Confidentiality ensures that measures are taken to prevent sensitive information from reaching the wrong persons. Data secrecy is important, especially for privacy-sensitive data such as user personal information and meter readings.

- *Nonrepudiation*: This is an assurance of the responsibility to an action. The source should not be able to deny having sent a message, while the destination should not deny having received it. This security objective is essential for accountability and liability.

Everybody is at risk for a cyber-attack. Cyber-attacks vary from illegal crimes of individual citizens (hacking) to actions of groups (terrorists). The following are typical examples of cyber-attacks or threats [9]:

- *Malware*: This is a malicious software or code that includes traditional computer viruses, computer worms, and

Trojan horse programs. Malware can infiltrate your network through the Internet, downloads, attachments, email, social media, and other platforms. Spyware is a type of malware that collects information without the victim's knowledge.

• *Phishing*: Criminals trick victims into handing over their personal information such as online passwords, social security numbers, and credit card numbers.

• *Denial-of-Service Attacks*: These are designed to make a network resource unavailable to its intended users. These can prevent the user from accessing email, websites, online accounts or other services.

• *Social Engineering Attacks*: A cybercriminal attempts to trick users into disclosing sensitive information. A social engineer aims to convince a user through impersonation to disclose secrets such as passwords, card numbers, or social security numbers.

• *Man-In-the-Middle Attack*: This is a cyberattack where a malicious attacker secretly inserts him/herself into a conversation between two parties who believe they are directly communicating with each other. A common example of man-in-the-middle attacks is eavesdropping. The goal of such an attack is to steal personal information.

These and other cyberattacks or threats are shown in Figure 10.4 [10].

Figure 10.4 Common types of cybersecurity threats [10].

The social and financial importance of cybersecurity is increasingly being recognized by businesses, organizations, and governments. Cybersecurity involves reducing the risk of cyberattacks. Cyber risks should be managed proactively by the management. Cybersecurity technologies such as firewalls are widely available [11]. Cybersecurity is the joint responsibility of all relevant stakeholders including government, business, infrastructure owners, and users. Cybersecurity experts have shown that passwords are highly vulnerable to cyber threats, compromising personal data, credit card records, and even social security numbers. Governments and international organizations play a key role in cybersecurity issues. Securing the cyberspace is of high priority to the US Department of Homeland Security (DHS). Vendors that offer mobile security solutions include Zimperium, MobileIron Skycure, Lookout, and Wandera.

10.3 CYBERSECURITY IN OIL AND GAS INDUSTRY

The oil and gas industry relies on complex technological systems to facilitate worldwide operations, making it highly vulnerable to cyber threats. Imagine the vast scope of the oil and gas industry: millions of miles of pipes, tankers traveling between ports, hundreds of refineries, rigs, production sites, and headquarters. The oil and gas supply chain is a globally interconnected environment, moving millions of barrels of crude oil and billions of cubic feet of natural gas on a daily basis. The complexity of the O&G systems is typically demonstrated in Figure 10.5 [12]. The oil and gas infrastructure is a tightly-knit network of extraction, production, refining, and distribution. The oil and gas industry is a prime target for cyber-attacks because of the value of the data and systems it controls. Any disruption can have major socioeconomic consequences due to cyber threats ranging from a data leak to tampering with control systems. Cyber breaches can compromise safety systems, leading to accidents, injuries, and potential environmental disasters.

Figure 10.5 The complexity of the O&G systems [12].

Five common cyberattacks on the oil and gas industry include [1]:

1.   *Phishing*: This type of attack involves sending emails or messages that appear to come from a legitimate source.

2.   *Ransomware*: This type of attack involves malware that encrypts the victim's files and demands payment or ransom in exchange for the decryption key.

3.   *Advanced Persistent Threats (APTs):* These are long-term, targeted attacks often conducted by nation-states or other highly skilled actors.

4.   *Distributed Denial of Service (DDoS) Attacks:* These attacks involve overwhelming a website or network with traffic to make it unavailable to legitimate users.

5.   *Industrial Control Systems (ICS) Attacks*: These attacks target the control systems that operate industrial processes, such as those used in oil and gas production.

The O&G industry faces many challenges in the service sectors of exploration and production. It is segmented into (upstream); processing, storage and transport (midstream); refining and processing (downstream).

- *Upstream Segment*: Upstream stages (exploration, development, and production and abandonment) have a distinct cyber vulnerability, and severity profile. Among the upstream operations, development drilling and production have the highest cyber risk profiles. The oil and gas production operation ranks highest on cyber vulnerability in upstream operations, mainly because of its legacy asset base, which was not built for cybersecurity but has been retrofitted and patched in bits and pieces over the years. A holistic risk management program could not only mitigate cyber risks for the most vulnerable operations but also enable all three of an upstream company's operational imperatives: safety of people, reliability of operations, and creation of new value. Apart from the upstream industry's "critical infrastructure" status, a complex ecosystem of computation, networking, and physical operational processes spread around the world makes the industry highly vulnerable to cyber-attacks.

Figure 10.6 shows cyber vulnerability by upstream operations [13].

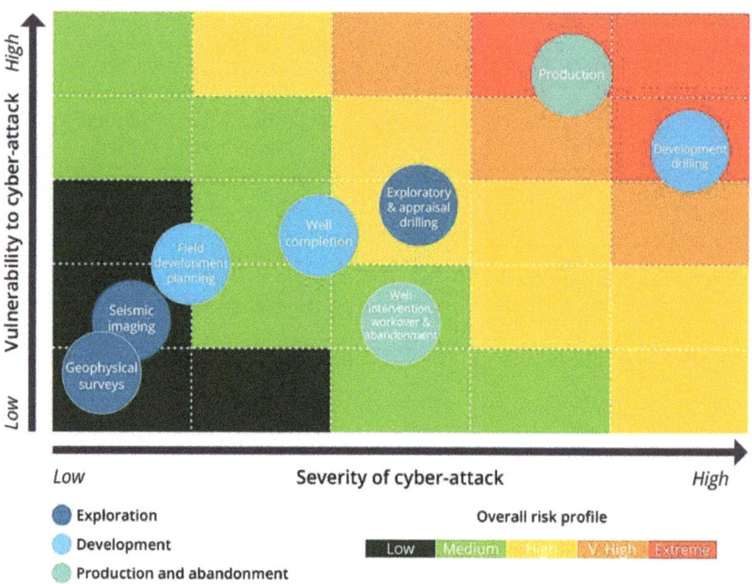

Figure 10.6 Cyber vulnerability by upstream operations [13].

- *Downstream Segment:* Cyberattacks can specifically impact the company's downstream business, which includes refining, chemical production, and distribution of petroleum products. Implementation of new technical and operational solutions in upstream and downstream processes was identified as the way forward towards digital oil fields since the early 2000s. For example, in downstream refineries, OT updates typically work around outage schedules so changes take much longer to implement in the oil & gas industry since facilities often operate 24/7.

10.4 ROLES OF CYBERSECURITY IN OIL AND GAS

As critical national infrastructure, oil and gas companies are key targets for cybercriminals. Cyberattacks are becoming more and more frequent and more sophisticated. These attacks are not only planned to disrupt operations but also to cause physical damage threatening human lives. Cybercriminals continue to look for new and innovative ways to infiltrate organizations. Their most common motive is to make money. Other motives of hackers range from cyberterrorism to industry espionage to disrupting operations to stealing field data.

Cyberattacks can compromise the availability, integrity, and confidentiality of oil and gas companies. They can endanger lives. Information systems (IT) and operational technology (OT) are also at risk. Their protection is crucial since it ensures the safety of people, systems, and data. The roles of oil and gas cybersecurity can be summarized as follows [14]:

- *Insider Threats*: Cyber threats are not strictly external; internal threats pose a significant risk as well. Unauthorized physical access to critical infrastructure can result in tampering or destruction of systems. Other so-called insider threats pose a significant challenge, as disgruntled employees, contractors, or others who have been granted prior authorized access can intentionally or unintentionally compromise critical systems and data. Cybercriminals regularly target workers with deceptive emails (i.e., phishing), compromising their credentials or tricking them into installing malware.

• *Supply Chain Risks:* Oil and gas professionals rank inadequate oversight of the vulnerabilities of supply chain partners connected to their organization's environment as the greatest challenge in enhancing OT cybersecurity. A supply chain attack on an oil and gas company is a cyberattack that targets the company's suppliers, vendors, or other partners to gain access to the company's systems and sensitive information. This can be done by compromising the security of a supplier or vendor and then using that access to move deeper into the company's network. The interconnected nature of the oil and gas industry introduces weaknesses through third-party vendors and suppliers. Those armed with privileged access can exploit vulnerabilities, compromise systems, or inadvertently expose critical information.

• *Layered Defense:* There is a need for active defenses and layers of protection for their most important assets. Companies can implement layers of protection for their most important assets. A layered defense, also called "defense in depth," is a proven concept based on various types of overlapping cybersecurity controls. The idea is that if one control fails or gets bypassed by the attacker, another layer offers protection. Companies can also implement active defenses like cyber intelligence, vulnerability awareness, and asset monitoring. Understanding the primary cybersecurity threats comes as the first step in building a robust defense.

- *Data Security:* Security-related data naturally claims an extremely high degree of sensitivity. This has resulted in major practical limitations in sharing data related to historical events and incidents related to cybersecurity in most cases. Oil and gas companies store large amounts of sensitive data, including intellectual property, financial records, exploration data, and customer information. Protecting this data is essential to prevent financial loss, reputational damage, and regulatory non-compliance. Sensitive data related to operations, production, and financial information needs to be protected from breaches, as this could expose valuable trade secrets and impact market competitiveness.

- *Encryption:* End-to-end encryption on all devices should be required and include embedded security. In some cases, certificate pinning must be required to avoid spoofed devices and this includes protection from side channel attacks that can compromise encryption keys.

10.5 BENEFITS

Companies in the oil and gas industry are attractive targets to cybercriminals because energy infrastructure is critical to modern economies. The disruption of an oil or gas pipeline has dire ripple effects on fuel prices, supply chains, and large-scale manufacturing. Building cyber resilience for your organization refers to its ability to survive and protect itself from a cyber-attack. By embracing these best practices, oil and gas companies will be well-positioned to fortify their cyber defenses, securing their critical infrastructure from would-be attackers. Energy companies must take strict measures to counter security threats. Benefits of these measures include the following [15]:

- *Automation*: Automate routine security tasks and orchestrate responses to reduce the burden on security teams, enabling them to focus on critical incidents and respond more efficiently. Implementing a security information and event management system to consolidate and correlate security events across your organization helps to significantly streamline alert management. Implementing automation and machine learning will filter alerts based on their severity and relevance. Your team is then able to analyze them in real-time, reduce false positives, and prioritize critical alerts for immediate attention.

- *Safety:* Increasing focus on safety integrity and the need for new preventative measures to safeguard against new safety risks have resulted in some new interests lately in both fault and failure diagnosis processes related to safety critical systems and equipment of offshore assets.

- *Security:* In the industrial world, productivity drives business. Today, there are security devices with industrialized hardware and advanced configuration options that can provide a defense-in-depth option for critical applications. These devices have sophisticated security capabilities including firewalls with integrated routers and VPN. Leading the charge in the oil and gas security and service market are major players such as Cisco Systems Inc., Honeywell International Inc., and Siemens. To avoid disruption of services and serious consequences, it is important to take a proactive stance on security, which starts with mitigating the risk of insignificant issues, long before they become major incidents. Companies should try to implement a platform that integrates physical security (CCTV, access control) with network security (firewalls, intrusion detection) and edge device management. This allows for continuous monitoring and detection of anomalous activity across all layers.

- *Monitoring:* For oil and gas organizations, modern video surveillance, which can be installed at any location, is required to better protect employees, assets, and the natural environment. Ideally, video surveillance is integrated with systems, such as access control, video analytics, and intrusion detection to detect threats faster. It is also important to deploy high-definition IP cameras and thermal imaging cameras to detect what can go unnoticed by regular surveillance.

10.6 CHALLENGES

The oil and gas industry faces significant cybersecurity challenges due to its reliance on increasingly interconnected IT and OT systems. Addressing these challenges and enhancing oil and gas cybersecurity is crucial to safeguard critical infrastructure, protect intellectual property, ensure safety, ensure business continuity, prevent cyber threats, and manage supply chain risk. Although emerging technologies will have a major impact on cybersecurity, they can be used for good and also for bad. Other challenges include the following [14]:

- *Worker Awareness:* Knowledge lays the foundation for secure processes and successful access management, and it also raises the awareness of all personnel. Raising oil and gas cybersecurity awareness among your staff helps foster a culture of safety, ensuring everyone understands their roles and responsibilities. With regular training programs, everyone should be periodically tested to eliminate complacency and inertia.

- *Cybersecurity Culture:* A new cybersecurity culture is an important industrial need that should be developed based on central attributes from different domains. The need for such a well-cultivated cybersecurity culture is immediate in all high-risk industrial sectors, such as the offshore O&G industry. Threats are constantly evolving and administering a culture of continuous improvement is one key step toward adopting a cybersecurity strategy that really works. Human factor is often considered as a weak point in the cybersecurity domain and hence the traditional approach of establishing cybersecurity culture often has a focus on raising human and organizational alertness.

- *Reducing Alert Fatigue*: Security alert fatigue is a common challenge for organizations juggling a large volume of warnings. In the oil and gas industry, reducing alert fatigue is an important factor in ensuring legitimate threats are promptly identified and addressed. Reduce alert fatigue by educating all workers, including upper management, about best practices for oil and gas security, thereby raising awareness about the potential consequences of being inattentive to the wide variety of cyber threats your organization regularly faces.

- *Standard:* Oil and gas companies need to constantly raise their standards. They must ask themselves whether they are learning and whether they are doing the right things. This internal scrutiny can be a source of confidence for companies. In 2018, the National Institute of Standards and Technology (NIST) proposed a cybersecurity framework (CSF) consisting of five steps: identify, protect, detect, respond and recover from a security incident. The CSF approach provides an effective framework for the simple integration of various standards, guidelines, and practices within the domains of industrial asset management and cyber risk management. Protecting control systems will be an ongoing responsibility that everyone must share by following standards and implementing a defense-in-depth approach to cybersecurity.

- *Regulation:* Regulatory requirements are the foremost driver of investment in cybersecurity within the oil and gas sector and the wider energy industry.

- *Collaboration*: It is paramount that oil and gas companies stay one step ahead of potential risks. Critical to staying ahead is fostering a culture of industry-wide collaboration. Frequently exchanging threat intelligence with international bodies and developing robust cybersecurity defenses can effectively mitigate cyber risks. To create a secure network, control engineers and plant managers must work together with the IT department and the technology they use.

10.7 CONCLUSION

The oil and gas companies play a crucial role in a functional, modern society. The companies face cyber threats daily, from hydrocarbon installation terrorism to industrial espionage. They are also vulnerable to theft and sabotage. Cybersecurity is paramount for the oil and gas industry as it plays a critical role in securing sensitive data. Ensuring the integrity, confidentiality, and availability of sensitive data and systems is of utmost importance. Cybersecurity of industrial control systems is multidisciplinary and involves multiple stakeholders, including operator companies, vendors, service providers, authorities, adversaries, and sometimes the public. Holistic cybersecurity ensures people, systems, critical infrastructure, and data are kept safe from cyberattacks.

Oil and gas companies that maintain cybersecurity as a central tenet of their digital strategy stand to gain the most. Emphasizing cybersecurity today means a more secure and resilient oil and gas infrastructure tomorrow. The defense against cyberattacks in oil and gas will have to be taken more seriously in the future. More information on cybersecurity in O&G industry is available from the books in [16-19] and the following related journals:

- *Energy and AI*

- *The AI Journal*

REFERENCES

[1] "Cyber threats for the oil and gas industry,"

https://meriplex.com/cyber-threats-for-the-oil-and-gas-industry/

[2] G. Lewis, "Mideast oil & gas facilities could face cyber-related energy disruptions," November 2023,

https://www.darkreading.com/ics-ot-security/mideast-oil-gas-facilities-could-face-cyber-related-energy-disruptions

[3] M. N. O. Sadiku, P. A. Adekunte, and J. O. Sadiku, "Cybersecurity in oil and gas industry," *International Journal of Trend in Scientific Research and Development*, vol. 8, no. 5, September-October 2024, pp. 830-838.

[4] P. Singh, "A layered approach to cybersecurity: People, processes, and technology- explored & explained," July 2021,

https://www.linkedin.com/pulse/layered-approach-cybersecurity-people-processes-singh-casp-cisc-ces

[5] M. Loi et al., "Cybersecurity in health – disentangling value tensions," *Journal of Information, Communication and Ethics in Society,* June 2019,

https://www.emerald.com/insight/content/doi/10.1108/JICES-12-2018-0095/full/html

[6] M. Adams, "Unlocking the benefits of ethical hacking: The importance of ethical hackers in cybersecurity," April 2023,

https://www.businesstechweekly.com/cybersecurity/network-security/ethical-hacking/

[7] M. N. O. Sadiku, S. Alam, S. M. Musa, and C. M. Akujuobi, "A primer on cybersecurity," *International Journal of Advances in Scientific Research and Engineering,* vol. 3, no. 8, Sept. 2017, pp. 71-74.

[8] M. N. O. Sadiku, M. Tembely, and S. M. Musa, "Smart grid cybersecurity," *Journal of Multidisciplinary Engineering Science and Technology,* vol. 3, no. 9, September 2016, pp.5574-5576.

[9] "FCC Small Biz Cyber Planning Guide,"

https://transition.fcc.gov/cyber/cyberplanner.pdf

[10] "The 8 most common cybersecurity attacks to be aware of,"

https://edafio.com/blog/the-8-most-common-cybersecurity-attacks-to-be-aware-of/

[11] Y. Zhang, "Cybersecurity and reliability of electric power grids in an interdependent cyber-physical environment," *Doctoral Dissertation,* University of Toledo, 2015.

[12] "Strengthening cybersecurity in the oil and gas industry," September 2024,

https://www.weforum.org/impact/cyber-resilience-oil-and-gas/

[13] "Protecting the connected barrels: Cybersecurity for upstream oil and gas,"

https://www2.deloitte.com/content/dam/insights/us/articles/3960-connected-barrels/DUP_Protecting-the-connected-barrels.pdf

[14] "Why is cybersecurity important for oil and gas?" July 2024,

https://www.otorio.com/blog/why-is-cybersecurity-important-for-oil-and-gas/

[15] P. Murchland, "Cybersecurity challenges in the energy industry – Are you ready?" July 2024, Unknown Source.

[16] M. N. O. Sadiku, *Cybersecurity and Its Applications.* Moldova, Europe: Lambert Academic Publishing, 2023.

[17] D. J. Lester, *Securing Oil and Natural Gas Infrastructures in the New Economy.* National Petroleum Council, 2001.

[18] C. Sundararaman, *Development of Cybersecurity Mandate For Oil And Gas Industry.* Self Publisher, 2023.

[19] A. Lamba, *Protecting 'Cybersecurity & Resiliency' of Nation's Critical Infrastructure - Energy, Oil & Gas*. SSRN, 2019.

275

INDEX

W

Waste, 234

Waste Management, 189

Water Management, 163

Water pollution, 184

Water treatment, 184

Wettability, 186